Essays In Pastoral Medicine

1906

JAMES JOSEPH WALSH

TABLE OF CONTENTS

PREFACE

The term Pastoral Medicine is somewhat difficult to define because it comprises unrelated material ranging from disinfection to foeticide. It presents that part of medicine which is of import to a pastor in his cure, and those divisions of ethics and moral theology which concern a physician in his practice. It sets forth facts and principles whereby the physician himself or his pastor may direct the operator's conscience whenever medicine takes on a moral quality, and it also explains to the pastor, who must often minister to a mind diseased, certain medical truths which will soften harsh judgments, and other facts, which may be indifferent morally but which assist him in the proper conduct of his work, especially as an educator. Pastoral medicine is not to be confused with the code of rules commonly called medical ethics.

The material of pastoral medicine requires constantly renewed discussion, because medicine in general is progressive enough frequently to devise better methods of diagnosis and treatment, and thus the postulates of the moral questions involved are changed. This discussion, however, is not easily made. The facts upon which the ethical part of pastoral medicine rests are furnished by the physician for the consideration and judgment of the moralist—the physician educated after modern methods knows little or nothing of ethics and can not himself make accurate moral decisions. The moralist, on the other hand, is commonly a poor counsellor to the physician, because long training in medicine is needed before the physical data of the moral decisions is comprehended. The physician, therefore, is at a loss to determine what he may or may not do in cases that involve the greatest moral responsibility, and the priest is a hesitating guide because the moral theologies do not convincingly present the doctrine in these cases.

Now and then such subjects have been proposed for discussion to a group of physicians and moralists, but usually no practical conclusion has been reached because one side did not understand the other. In 1898 there was a series of articles on ectopic gestation in the American Ecclesiastical Review, in which moralists like Lehmkuhl, Sabetti, Aertnys, and Holaind, and some of the leading gynaecologists of America considered the questions but arrived at no decision. The physicians did not understand certain questions, other questions were on obsolete medical practice, essential questions were omitted, and from the data the moralists came to opposed conclusions.

We find also in moral theologies deductions drawn from false medical sources. Reasons are given, for example, to justify the use of a large quantity of alcoholic liquor at a dose in cases of great pain, typhoid fever, snake-poisoning, and other diseases, in the supposition that such doses will benefit or cure the patient, whereas the physician that would follow that treatment would be guilty of malpractice. There was recently in America a discussion on the relation of öophorectomy to the impedimentum impotentiae. One side held that a lack of ovaries constitutes impotence; the other side, that it does not. The discussion was useful because it incidentally gathered the full doctrine of the moralists on this subject, but from the medical point of view there is no connection whatever between these conditions.

A small number of books on pastoral medicine have been written by clergymen that were not physicians, and a few German books by physicians that were also moralists. Those by the physicians draw conclusions from antiquated medical practice, or they are mere popular treatises on hygiene; those by the clergymen have some value on the ethical side, but they are incomplete because the authors had not the necessary medical knowledge. The essays offered in this book by no means cover the entire field of pastoral medicine, but as far as they go we have endeavoured to offer the medical doctrine of the present day on the questions considered, and that as completely as is necessary to draw the moral inferences.

Since, then, so many of the questions of pastoral medicine are not defined, physicians are likely to follow the doctrine of the standard medical books, which without exception advise them to take the life of a dangerous foetus almost as unconcernedly as they might prescribe an active drug, or in any case to put utility before justice. There is, therefore, an urgent necessity that competent men fix that shifting part of ethics and moral theology called pastoral medicine, and these essays are presented as a temporary bridge to serve in crossing a corner of the bog until better engineers lay down a permanent causeway.

Some may think that the authors are inclined toward an exaggerated charity in suggesting the measure of responsibility for many human actions, but the physician that is brought much in contact with those suffering from mental defects of various kinds soon learns how easily complete responsibility becomes marred. Responsibility is dependent entirely upon free will; and while the great principle of free will remains solid in truth, no two men are free in exactly the same manner. Physical conditions have not a little to do with modifications of freedom of the will. To point out this fact to the clergyman and the physician has been our intention, for a proper appreciation of it will widen the bounds of charity and save many that are more sinned against than sinning from the injury of grievous misjudgment. It is better to run the risk of exculpating a few individuals whose responsibility is not entirely clear when the application of the same principles lifts many others above the rash judgment of those that can be of most help to them.

ECTOPIC GESTATION

Ectopic gestation is gestation in the uterine adnexa, the peritoneal cavity, or the horn of an abnormal or rudimentary uterus. It is opposed to natural uterine gestation, and, since it includes pregnancy in an abnormal uterus, it is a more comprehensive term than extrauterine pregnancy.

In this article the morality involved in the surgical treatment of ectopic gestation is considered; and to have the data requisite for judgment it is necessary to describe in outline the anatomy of the uterine adnexa and the growth of the foetus; to explain the varieties, effects, diagnosis, and treatment of ectopic gestation; to present the cases of this condition, or rather this disease, as they occur in medical practice; to set forth some of the moral principles or laws that govern medical practice, especially where there is question of life and death; and finally to apply these principles to the cases offered for investigation.

The uterus is in the pelvic cavity, between the bladder and the rectum and above the vagina, into which it opens. It is a hollow, pear-shaped, muscular organ, somewhat flattened, and about three inches long, two inches broad, and one inch thick. The base or fundus is upward, and the neck is downward. Passing out horizontally from the corners or horns of the uterus, which are at its base, are the two Fallopian Tubes, one on either side. These are about five inches in length and somewhat convoluted. They are true tubes, opening into the uterus, and they are about one-sixteenth of an inch in diameter along the greater part of their extent The ends farthest from the uterus are fringed and funnel-shaped; and this funnel-end, called the Infundibulum or the Fimbriated Extremity, opens into the abdominal or peritoneal cavity. Near the Fimbriated Extremity of each tube is an Ovary,—an oval body about one and a half inches long by three-quarters of an inch in width.

The Uterus and its Adnexa

F U, Fundus or Base of the Uterus. F T, P T, Fallopian Tubes. On the left of the reader the Fimbriated Extremity of the tube is lifted up to show it. O, O, Ovaries. B L, B L, Broad Ligament. R, Rectum. B, Bladder.

For convenience in description, each tube is divided into four parts: (1) the Uterine Portion, which is that part included in the wall of the uterus itself; it extends from the outer end of the horn into the upper angle of the uterine cavity, and its lumen is so small it will admit only a very fine probe; (2) the Isthmus, or the narrow part of the tube which lies nearest the uterus; it gradually opens into the wider part called (3) the Ampulla; (4) the Infundibulum, or the funnel-shaped end of the Ampulla. This part is fimbriated, as has been said, and one of the fimbriae—the Fimbria Ovarica—which is longer than the others, forms a shallow gutter which extends to the ovary.

The uterus, tubes, and ovaries lie in a septum which reaches across the pelvis from hip to hip. This septum is called the Broad Ligament. If a man's soft felt hat, of the kind called a "Fedora" hat, is held crown downward with one hand at the front and the other at the back of the rim, it will represent the pelvic cavity, and the fold along the crown of the hat coming up into this cavity is very like the Broad Ligament. As the crown is held downward, the uterus would be in the middle, its fundus upward, and, of course, altogether outside the hat, but in the crown fold. The tubes and ovaries would also be outside the hat and in the crown fold, and the fimbriated extremities would open by holes into the hat's interior.

The ovum breaks through the surface of the ovary, passes, probably on a capillary layer of fluid, into the fimbriated extremity of the tube, and then is moved along slowly through the tube into the uterus. Ovulation and menstruation occur about the same time, but often one antedates the other a few days. In exceptional cases they may occur independently.

If the ovum produced is not fecundated, it gradually shrivels up, and passes off through the uterus and the vagina. Fecundation of the ovum rarely occurs in the uterus, but ordinarily in the Fallopian tube, according to the general opinion of physiologists. After fecundation the ovum is pushed on into the uterus in from five to seven days, where it fastens to the wall and develops. Hyrtl (Kollmann's Lehrbuch der Entwickelungsgeschichte des Menschen, Jena, 1898) speaks of a case in which the ovum appeared to reach the uterus in three days. If the fecundated ovum is blocked or held in the Fallopian tube, the embryo grows where the ovum stops, and we have a case of Ectopic Gestation.

The average time of normal human gestation is ten lunar months or forty weeks. At the moment the pronucleus of the spermatozoon fuses with the pronucleus of the ovum in the Fallopian tube and makes the segmentation nucleus, in my opinion, the soul of the child enters, and personality exists as absolutely as it does in a child after birth. It is as much a murder, as such, to unjustly destroy this microscopic fecundated ovum as it is to kill the child after birth. This is the opinion of every embryologist I have consulted on the subject, with the exception of one who said he did not know when the soul enters.

Technically the product of conception is called the Ovum for the first two weeks of pregnancy; during the third and fourth weeks it is called the Embryo, and after the fifth week the Foetus. During the fourth week the embryo begins to draw nourishment from the maternal blood through its umbilical vessels, but before that time it obtains nourishment by osmosis.

The foetus at the end of the eighth week is about one inch in length; at the end of the fourth lunar month it is from four to six inches long, and its sex may be distinguished. At the end of twenty-four weeks, if the normal foetus is born it will attempt to breathe and to move its limbs, but it dies in a short time. At the end of twenty-eight weeks of gestation if it is born it moves its limbs freely and cries weakly. It is nearly fifteen inches in length and weighs about three pounds. Such an infant might be deemed viable, but its chances for life are extremely precarious, even in most expert hands and with the help of an incubator. At the end of thirty-two weeks of gestation a foetus if born may be raised with skilful care, but the chances are not promising. It is viable. At the end of forty weeks the child is at term.

In 1876 Parry collected 500 cases of extrauterine pregnancy from medical literature, but when Tait in 1883 first operated on a case of ruptured tubal pregnancy attention was called to the subject. It was better understood as coeliotomies (opening the abdomen) became common, and in 1892 Schrenck collected 610 cases that had been reported during the preceding five years. Küstner alone has operated on 105 cases in five years.

There has been much discussion among physicians as to the causes that arrest the fertilized ovum in the tube, but whatever these causes may be they do not affect the moral questions which come up in this article. There may be mechanical obstruction from peritoneal adhesions, or abnormal conditions resulting from inflammatory diseases of the tubes, ovaries, and the pelvic peritoneum, but no general cause that will explain all cases can be ascribed.

Tait denied the possibility of Ovarian Pregnancy, or a pregnancy where the ovum fastened to the ovary itself and developed there, but five fully established cases of this kind have been reported. Dr. J. Whitridge Williams, professor of Obstetrics at the Johns Hopkins University, in his textbook on Obstetrics (New York, 1903), collects twenty-five cases of ovarian pregnancy, where five cases are certain diagnoses, thirteen highly probable, and seven fairly probable. In these twenty-five cases ten foetuses reached full term, but four of the five certain cases ruptured at early periods.

It was formerly thought that primary abdominal pregnancy is quite common; that is, that the ovum is implanted on some organ within the abdomen itself, apart from the uterine adnexa. This is now looked upon as very doubtful, and such cases are probably secondary; that is, secondary to a pre-existing tubal pregnancy which has ruptured without great maternal hemorrhage and let the foetus grow within the peritoneal cavity.

The common form of extrauterine pregnancy is the Tubal Pregnancy. The ovum may be stopped in any one of the three parts of the tube, and we find Interstitial, Isthmic, or Ampullar Pregnancy. From these primary types, by rupture, secondary forms sometimes arise,—Tubo-abdominal, Tubo-ovarian, and Broad-ligament Pregnancy.

The interstitial form, that is, where the ovum is arrested in that part of the tube which passes through the wall of the uterus itself, is the rarest of the tubal pregnancies. Rosenthal (Ein Fall intranturaler Schwangerschaft. Centralbl. f. Gyn. 1297-1305) found it in only three per centum of 1324 cases of tubal pregnancy. Some deem the Isthmic variety the commonest. Dr. Howard Kelly (Operative Gynaecology) says he never met a case of Interstitial or Ovarian pregnancy in his practice. The interstitial form is especially liable to rupture with suddenly fatal hemorrhage.

About one-fourth of the cases of tubal pregnancy end within the first twelve weeks by rupture of the Fallopian tube. If the embryo is implanted in the interstitial end of the tube, the rupture (into the uterus, or into the abdominal cavity, or into the broad ligament) takes place later,—about the fourth month, or even considerably after that time. The reason for the delay here is that the uterus grows with the foetus. If the foetus breaks into the uterus (a very rare occurrence), it is either expelled through the vagina almost immediately or it goes on like a normal pregnancy.

Tait was of the opinion that every case of tubal pregnancy results in a rupture of the tube not later than the twelfth week, but this opinion is no longer held. Very rarely a tubal pregnancy goes on without rupture to full term, as in the cases reported by Williams, Saxtorph, Spiegelberg, Chiari, and a few others.

Three-fourths, about seventy-eight per centum, of the cases of tubal pregnancy result in what is technically called "tubal abortion" instead of rupture. In tubal abortion the connection between the embryo and the tube-wall is broken by effusion of blood. If the separation is complete the effused blood pushes the embryo out through the fimbriated end of the tube into the abdominal cavity, and then the hemorrhage of the mother commonly ceases. Such an extrusion of the foetus is called a complete tubal abortion. If the connection between the foetus and the tube-wall is only partly severed, the ovum remains in the tube, and the maternal hemorrhage goes on. This is called incomplete tubal abortion.

In incomplete tubal abortion the maternal blood may slowly trickle from the fimbriated extremity of the tube into the abdominal cavity, become encapsulated, and thus form an haematocele. If the fimbriated extremity of the tube is blocked, the blood accumulates in the tube and makes an haematosalpynx.

In complete tubal abortion the foetus dies; in incomplete tubal abortion the viability might depend on the injury done the placenta, but in almost every case of even incomplete tubal abortion the foetus dies as a result of its separation from the tubal wall, or from compression after the bleeding.

In cases of rupture of the tube in extrauterine pregnancy, if the foetus with its attachments is expelled from the tube into the peritoneal cavity or into the broad ligament, the embryo dies.

If the foetus or embryo itself alone is expelled into the abdominal cavity and the placenta remains attached to the wall of the tube and communicates with the foetus by the umbilical cord which runs through the tear in the tube, the foetus may possibly live, provided the mother does not die from hemorrhage. If the foetus goes on growing in this case, we have an abdominal pregnancy. One such case is reported by Both where a fully developed foetus was found in the abdominal cavity even lacking all its membranes, which had been left in the tube, but a foetus will not live apart from its membranes within the maternal body.

When an embryo or foetus ruptures the tube and goes into the broad ligament, it may live or die according to the injury done its attachments to the tubal wall, but it ordinarily dies. Sometimes such a broad-ligament pregnancy ruptures again into the abdominal cavity. Because the bleeding is more likely to be confined within the folds of the broad ligament, the immediate danger of maternal death from hemorrhage is less in this than in other forms of rupture.

Concerning tubo-abdominal pregnancy the only remark to be made is that, owing to adhesions, it is often surgically difficult to remove such a growth.

If the foetus is expelled after rupture into the peritoneal cavity it dies, and if the hemorrhage does not kill the mother the dead foetus if small is absorbed; if large it becomes mummified, or it hardens into a lithopoedion, or it turns into a yellowish greasy mass called adipocere, or it putrefies. A lithopoedion may be carried for years. There are more than thirty cases reported which were carried from twenty to thirty years in the abdomen, and one case where a lithopoedion was carried for fifty years.

If the foetus putrefies it causes fatal septicaemia in the mother, or a perforating abscess, unless it is successfully removed.

There are various abnormalities of the uterus, and in these pregnancy resembles in effect extrauterine pregnancy. An abnormal uterus may be unicornis, didelphys, pseudodidelphys, bicornis duplex, bicornis septus, bicornis subseptus, bicornis unicollis, or bicornis unicollis with a rudimentary horn. The impregnated ovum may fasten in the rudimentary horn and be blocked there; then the usual result is rupture within the first four months, with fatal hemorrhage unless the bleeding is immediately checked by coeliotomy and ligation.

As to diagnosis in Ectopic Gestation, Williams (op. cit.), one of the authorities at present on the subject, says: "A positive diagnosis is occasionally made before rupture, but in the vast majority of cases the condition escapes recognition until symptoms of collapse point to the probability of rupture or abortion. In advanced cases careful examination will usually disclose the real condition of affairs, and when full term has been passed the history is so characteristic that mistakes should hardly occur."

In the American Ecclesiastical Review for January, 1898 (vol. ix., n. i), Father René I. Holaind, S. J., published the answers of many physicians to six questions concerning extrauterine pregnancy. Among these physicians were Thomas Addis Emmet, Barton Cooke Hirst, Howard A. Kelly, W. T. Lusk, T. Galliard Thomas, Mordecai Price and his brother Joseph Price, William Goodell, and Lawson Tait,—all eminent authorities on this subject. The second question submitted was: "During pregnancy, at what time and by what means can a differential diagnosis be made between intra and extrauterine pregnancy, and between abnormal gestation and pelvic or other tumour?"

In answer to this question Dr. Emmet said: "There can be no absolute certainty as to the existence of pregnancy in any case until the pulsation of the foetal heart can be detected. [After the eighteenth or twentieth week of gestation.] ... A diagnosis is difficult in all cases of abnormal pregnancy, but an expert can, within a reasonable degree of certainty, arrive at a knowledge of the existing conditions between the second and third month."

Dr. Hirst said: "In almost all cases of advanced gestation the differential diagnosis can be made. In early cases it is not always possible unless conditions be favourable."

Dr. Howard A. Kelly said: "The differential diagnosis between intra and extrauterine pregnancy can usually be made from the sixth week up to the end of pregnancy. It is more easily made from the tenth to the twelfth week on." Writing in the American Text Book of Obstetrics (Philadelphia, 1896), he says: "In the atypical cases, on the contrary, a positive diagnosis is often difficult or even impossible. ... The diagnosis of ectopic gestation after the death of the foetus is largely dependent upon the clinical history; if this be deficient, the diagnosis is frequently impossible."

Dr. Lusk said: " ... The frequent discovery of the dead ovum in a tube when there has been no suspicion of pregnancy shows the difficulty of a diagnosis." In his text-book (The Science and Art of Midwifery, New York, 1890) is this remark: "Sometimes the diagnosis can only be decided by the introduction of the sound or a finger into the uterus, the physician assuming the risk of premature labour, should he find his supposition of extrauterine pregnancy an error." This means that sometimes the diagnosis is impossible without running the risk of causing abortion of a normal uterine pregnancy.

Dr. Thomas said, "After the second month the diagnosis is perfectly possible." This was also the opinion of Dr. Mordecai Price; and Dr. Joseph Price holds that the diagnosis can be made "after the third month, by exclusion." Dr. John F. Roderer, quoting Lawson Tait, says that "the diagnosis between intra and extrauterine pregnancy can not be made with certainty before rupture, nor can it be determined exactly whether an enlargement of the tube is either an ectopic pregnancy or some form of tumour."

Dr. Goodell's opinion was, "A differential diagnosis can rarely be made positively at any stage of extrauterine pregnancy."

The diagnosis, then, is difficult; and for the ordinary practitioner, the average physician, who does perhaps ninety-five per centum of the medical work of the world, the diagnosis is often impossible. There is no greater expert than Dr. Thomas Addis Emmet, and he says the diagnosis is difficult. Others hold that the diagnosis can be clearly made, and they speak truly as regards themselves, but ordinary skill finds the diagnosis almost impossible in many cases. Mordecai Price (The Pennsylvania Medical Journal, vol. viii. p. 223) in one year saw four cases which he and other physicians diagnosed as ectopic pregnancies with rupture of the tube. When the abdomen had been opened, uterine pregnancy was discovered with a ruptured tube in each case, and all the women died.

The first positive diagnosis of unruptured tubal pregnancy was made by Veit in 1883, and the first one made in America was by Janvrin in 1886, eight years before Father Holaind's article was written. Before 1883, only eleven years in advance of the same article, when Lawson Tait performed the first coeliotomy for the purpose of checking hemorrhage from a ruptured tubal gestation, extrauterine pregnancy was as mysterious as the old "inflammation of the bowels," which turned out afterward to be appendicitis. Hence common skill in the difficult diagnosis of ectopic gestation can not be looked for.

The doctrine given in all the leading medical works at present concerning the treatment of extrauterine pregnancy is this:

1. As soon as an extrauterine pregnancy is discovered remove the foetus through an opening made in the mother's abdominal wall. Do not use electricity or the injection of poisons into the foetal sac, or the incandescent knife. Emmet and a few others approved of the use of electricity at times, but this is against the teaching of the great majority of writers at present. The reason for removing the foetus at once is that it is apt at any moment to cause rupture and fatal hemorrhage before surgical aid can be effective.

2. In a case of rupture with free hemorrhage and collapse the only

operation advised is an immediate coeliotomy to stop the bleeding by ligatures. The rupture should not be approached through the vaginal wall according to the common doctrine, but through the abdominal wall.

3. If there is a rupture in which the bleeding is confined and there is no collapse, do not operate at once unless the haematocele increases steadily or shows signs of suppuration. Sometimes evacuation of the haematocele through the vaginal wall is possible.

4. In the later months of an extrauterine pregnancy, whether the case is intraligamentous or abdominal, perform coeliotomy as soon as the diagnosis has been made, and remove the foetus, because there is always danger of sudden and fatal hemorrhage before the surgeon can reach the source of the bleeding. What is to be done in a case where the surgeon is certain before operating that the foetus is dead, has interest only for the physician, and it involves no moral question.

Operating for extrauterine pregnancy maybe a simple coeliotomy, if any coeliotomy is really simple, but it commonly is the most dangerous operation for the mother that the gynaecologist is called upon to perform.

The discussion of the moral questions that arise in cases of ectopic gestation which began in volume ix. of the American Ecclesiastical Review was very valuable, but as the moralists had not full data to work on their decision as a whole is not satisfactory. The original cases presented are in part obsolete in the medical practice of to-day, and important physical conditions were not disclosed in some of the other parts of the cases. Father Holaind tentatively agreed with Father Lehmkuhl in one decision, Fathers Eschbach and Sabetti directly attacked Father Lehmkuhl's reasons, and Father Aertnys indirectly opposed Father Sabetti's chief argument. These men are all eminent authorities, but as each, except Father Holaind, was dissatisfied with the arguments advanced by the others, and as their data were incomplete, we can not rest the case on their decision.

In Father Holaind's fifth question, if I understand it correctly, he seemed to think it possible to baptize a foetus through the opening in the mother's abdominal wall while it lies in the abdominal cavity before surgical removal. He mentions antiseptic precautions in the baptism, which would have no meaning if the foetus were out of the abdomen.

Baptism would not be possible in that case: the priest could not get at the foetus, he ordinarily could not even see it, and certainly no surgeon would permit the attempt. There would be no time for the attempt in a rupture case, even if the foetus could be seen; and there would be no advantage gained by baptizing the foetus in the abdominal cavity where the conditions gave time to do so. If it is alive it will live long enough for baptism after removal from the abdomen, provided, of course, it is baptized immediately in the operating room. That it does not breathe is no proof of immediate death. It is not unusual for a full-term child not to breathe for even an hour or longer after birth.

If Father Holaind had not in view baptism within the abdominal cavity, the question has this meaning: What is the most effective method after the foetus has been removed from the abdomen to open its enveloping membranes so as to give it a chance for a life lasting long enough to allow baptism?

The best method is to slit the membranes with a scalpel or scissors as quickly as possible. The envelopes, cord, and placenta are essential parts of the foetus itself, and they grow from itself, not from the mother. They take the place of the lungs and the alimentary tract, which do not come into action until after normal birth. It would be worth discussing whether a baptism on the intact foetal envelopes is valid, were it not that we may not apply probabilism in such a case. The remaining matter brought out in Father Holaind's questions will be considered in the course of this article.

Before presenting the cases of ectopic gestation that occur in medical practice, the fundamental ethical principles that are to be applied in judging the morality of the surgeon's interference should be given.

The morality of any action is determined, (1) by the object of the action; (2) by the circumstances that accompany the action; (3) by the end the agent had in view.

1. The term object has various meanings, but here it means the deed performed in the action, the thing which the will chooses. That deed by its very nature may be good, or it may be bad, or it may be indifferent morally. In themselves to help the afflicted is a good action, to blaspheme is a bad action, to walk is an indifferent action. Some bad actions are absolutely bad, they never can become good or indifferent (blasphemy or adultery, for example); others, as stealing, are evil because of a lack of right in the agent: these may become good by acquiring the missing right. Others are evil because of the danger necessarily connected with their performance,—the danger of sin connected with them, or the unnecessary peril to life. An action to have the moral quality must be voluntary, deliberate; and mere repugnance in doing an act does not in itself make the act involuntary.

2. Circumstances sometimes, though not always, can add a new element of good or evil to an action. The circumstances of an action are the agent, the object, the place in which the action is done, the means used, the end in view, the method observed in using the means, the time in which the deed is done. If a judge in his official position tells a sheriff to hang a criminal, and a private citizen gives the same command, the actions are very different morally because of the circumstances of the agent giving the command. The object—it changes the morality of the deed if a man steals a cent or a thousand dollars. The place—what might be merely a filthy action in a house might be a sacrilege in a church. The means—to support a family by labour or by thievery. The end in view—to give alms in obedience to divine

command or to give them to buy votes. The method observed in using means—kindly, say, or cruelly. The time—to do manual labour on Sunday or on Monday. Some circumstances aggravate the evil in a deed, some extenuate it. Others may so colour a deed that they specify the deed, make the action some special virtue or vice. The circumstance that a murderer is the son of the man he kills specifies the deed as parricide.

The end also determines the morality of an action (see St Thomas, Sum. Theol. I. 2., q. xviii., a. 4 and 7). Since the end is the first thing in the intention of the agent, he passes from the object wished for in the end to choosing the means for obtaining it. Without the end the means can not exist as such. There are occasions when an end is only a circumstance: for example, if it is a concomitant end. When an end is a, finis extrinsecus operantis, when it is in keeping with right reason or discordant thereto, it may become a determinant of morality.

In every voluntary, or human, act there is an interior and an exterior act of the will, and each of these acts has its own object. The end is the proper object of the interior act of the will; the exterior object acted upon is the object of the exterior act of the will; and as this exterior act specifies the morality, so does the interior object—which is the end—specify it, and even more importantly than the exterior object does.

The will uses the body as an instrument on the external object, and the action of the body is connected with morality only through the will. We judge the morality of a blow, not by the physical stroke, but from the intention of the striker. The exterior object of the will is, in a way, the matter of the morality, and the interior object of the will, or the end, is the form. Aristotle (Ethics, lib. v., cap. 2) says: "He that steals that he may commit adultery, is, absolutely speaking, more an adulterer than a thief." The thievery is a means to the principal end, and it is this principal end that chiefly specifies or informs the action.

The means used to obtain an end are very important in a consideration of the morality of an act. There are four classes of means,—the good, the bad, the indifferent, and the excusable.

Good means may be absolutely good, but commonly they are liable to become vitiated by circumstances,—almsgiving is an example. Some means are bad always and inexcusable,—lying, for example. The excusable means are those which are bad, but justifiable through circumstances. To save a man's life by cutting off his leg is an excusable means.

The existence of excusable means whereby some good actions are effected does not establish the assertion that the end justifies the means. The end sometimes may incriminate or sanctify indifferent means, but it does not in itself justify all means. The means, like other circumstances, are accidents of an action, but they are in an action just as much as colour is in a man. Colour is not of a man's essence, but you can not have a man without colour.

The effect of an action, the result or product of an effective cause or agency, may in itself be an end or an object or a circumstance, and it has influence in the determination of morality. Sometimes an act has two effects, one good and the other bad; and that such an action be lawful it is necessary (1) that the action itself be good or indifferent; (2) that the good effect be intended and the evil effect be not intended (chosen) but only reluctantly permitted; (3) that the evil effect be not a means to secure the good effect; (4) that there be present a motive sufficiently grave to excuse or counterbalance the bad effect. St. Thomas (Sum. Theol. 2. 2. q. 64, a. 7) Speaking of killing a man in self-defence, says: "Nihil prohibet unius actus esse duos effectus, quorum alter solum sit in intentione, alius vero sit praeter intentionem. Morales autem actus recipiunt speciem secundum id quod intenditur, non autem ab eo quod est praeter intentionem, cum sit per accidens."

That an act, therefore, be morally good, or justifiable, (a) the whole train of the tendency of the will must be good; that is, (1) the object, (2) the end, (3) and the circumstances must be good; or (b) the intention should be good, and the remaining elements in the train of will-tendency are to be indifferent. That an act be morally bad it is enough that the object, the end, or the circumstances be inexcusably bad.

There may be honest doubt as to the existence of evil in the circumstances or the end, and here enters the matter of probability; but apart from this, some general rules of morality that govern all cases may be formulated:

1. An intention or end which is gravely evil always makes the entire action evil and unjustifiable.

2. An intention or end which is slightly evil, if it is the entire end of an action, makes the whole action evil but not gravely evil—makes it, say, a venial sin and not a mortal sin.

3. If an intention or end which is venially evil accompanies secondarily a good intention or end, and is rather a motive than the real effective agent in attracting the will, this venial evil does not vitiate the whole goodness or righteousness of the main action. Compare the remarks made above in discussing an action that has a double effect, partly good and partly bad.

4. Circumstances that are gravely evil practically vitiate the entire action, but circumstances which are venially evil do not always vitiate the entire action.

Much might be said here concerning conscience as a judge of the morality in an act, but this discussion is not necessary for our present purpose. Like other men, physicians often confuse conscience with inclination, or at best with unfounded opinion. When conscience is to be a rule of action it must have at the least moral certitude; or, what is different but practically the same thing, the opinion of conscience must be at the least genuinely probable. The term "probable" is used here in a technical sense, and it will be so used throughout the remainder of this article.

The doctrine of Probabilism is connected with the promulgation of law. A law, according to St. Thomas (op. cit. I. 2., q. 90, a. 4) is: "Ordinatio rationis ad bonum commune ab eo qui curam habet communitatis promulgata." Sometimes it is not evident whether or not a law binds in a particular case, and in such a condition, that is, in which there is question solely of the existence, interpretation, or application of a law, we may follow a probable opinion which assures us the act is licit, although the opinion which says the act is illicit may be just as probable or even more probable. This is the fundamental proposition of Probabilism, which is the doctrine especially of St. Alphonsus Liguori, but it was held centuries before his time. As the church has never condemned this doctrine, but rather tacitly approved of it, Catholics may safely follow it, and those that are not Catholics will find it very reasonable.

A law which is doubtful after honest and capable investigation has not been sufficiently promulgated, and therefore it can not impose a certain obligation because it lacks an essential element of a law. When we have used such moral diligence of inquiry as the gravity of a matter calls for, but still the applicability of the law is doubtful in the action in view, the law does not bind; and what a law does not forbid it leaves open.

Probabilism is not permissible when there is question of the worth of an action as compared with another, or of issues like the physical consequences of an act. If a physician knows a remedy for a disease that is certainly efficacious and another that is probably efficacious, he may not choose the probable cure, at the least in a grave illness. Probabilism has to do with the existence, interpretation, or applicability of a law, as I said, not with the differentiation of actions.

The term probable means provable, not guessed at, or jumped at without reason. There must be sound reason adduced to constitute probability. The doubt must be founded on a positive opinion against the existence, interpretation, or application of the law. It must be more than mere negative doubt, more than ignorance, more than vague suspicion, especially must it be more than a sentimental impression. There is a mental condition, which easily passes over into disease, wherein a man habitually can not make up his mind. This flabbiness has nothing to do with Probabilism. The opinion against a law to constitute Probabilism must be solid. It must rest upon an intrinsic reason from the nature of the case, or an extrinsic reason from authority,—always supposing the authority cited is really an authority. Many men sitting upon the supreme bench in the Court of Science and called authorities by friends and newspapers, are only fools in good company.

The probability must also be comparative. What seems to be a very good reason when standing alone may be very weak when compared with a reason on the other side. When we have weighed the arguments on both sides, and we still have good reason left for standing by our opinion, our opinion is probable. The probability is, moreover, to be practical. It must have considered all the circumstances of the case.

The principles presented here have been arranged, as we said, with a view toward application in judging the morality of actions that may occur in cases of ectopic gestation, and we shall apply the doctrine of probabilism in the question, does the commandment "Thou shalt not kill" bind in certain cases of ectopic pregnancy? It is also necessary to add the principles underlying our duty to preserve human life.

1. It is never lawful directly or indirectly to kill an innocent man. "Insontem et justum non occides" (Exod. xxiii. 7). An innocent man is one that has not by any human act done harm to another man or to society commensurate with the loss of his life. Directly means to kill either as an end, say, for revenge, or as a means toward an end.

A man is a person, an intelligent being, therefore free, and autocentric; he belongs to no one except to God, who made him; he is by that very fact distinguished from brutes or things which may belong to another. Now, if you kill a man, you destroy his human nature by separating his soul and body, you subordinate and sacrifice him wholly to yourself, make him entirely yours, which is unjust. Even the state has no right to kill an innocent man. A foetus in the womb, only a few hours old, is as much a human being as a man fifty years of age, and this natural law holds for the foetus as for the man.

2. It is, however, lawful indirectly to kill a man provided this man is an unjust aggressor. Cardinal de Lugo (De Just. et Jure, 10, 149) and others hold you may even directly kill an unjust aggressor. Indirectly here means incidentally. An effect happens indirectly when it is neither intended as an end nor a means, but happens as a circumstance unavoidably attached to the end or means intended.

We may not, however, kill an innocent man even indirectly, because no end is proportionate to the sacrifice of an innocent man's life, but the case of an unjust aggressor differs from that of an innocent man. By an unjust aggressor is meant some one that outside the due course of law threatens your life or the equivalent of your life, or the life of some one you should or may protect. You may stop such an aggressor, and if you happen to kill him while trying to stop him, there is no moral wrong involved. This aggressor may be formally or only materially unjust: he may be a normal man with a formal intention to kill you or your ward, or a murderous lunatic that tries to kill you or your ward, but he must be unjust either formally or materially.

It is natural for every being to maintain itself in existence, to resist destruction. This is a primary law of nature. As Father Holaind well said (Amer. Eccl. Rev., January, 1894): "The ethical foundation of self-defence is this: Justice requires a sort of moral equation, and if a right prevails it must be superior to the right which it holds in abeyance. At the outset both the aggressor and his intended victim have equal rights to life, but the fact of the former using his own life for the destruction of a fellow man places him in a condition of juridic inferiority with regard to the latter. If we may be allowed so to express it, the moral power of the aggressor is equal to his inborn right to life, less the unrighteous use which he makes of it, whilst the moral power of the intended victim remains in its integrity and has consequently a higher juridic value. When the person assailed cannot defend himself, his right can and sometimes must be exercised by those who are bound in justice or charity to protect the innocent. At the dawn of human life the physician or surgeon stands as the natural protector both of the mother and of the child; he is beholden to both.

"The right of self-defence is not annulled by the fact that the aggressor is irresponsible. The absence of knowledge saves him from moral guilt, but it does not alter the character of the act, considered objectively and in itself; it is yet an unjust aggression, and in the conflict, the life assailed has yet a superior juridic value. The right of killing in self-defence is not based on the ill will of the aggressor but on the illegitimate character of the aggression. Now, an aggressor is at least materially unjust whenever he perpetrates an act destructive of the right of another."

Mark the words "right of another," at the end of the quotation. In a case of pregnancy at term in a woman with a contracted pelvis the foetus would be a contributing instrument of death to the mother, supposing there were no artificial means of delivering her, but such a child is not an aggressor even materially unjust. The child itself is normal, it has a natural right to be where it is, it did not put itself where it is; the mother's contracting uterus crushing the child against her narrow pelvic arch is the direct agency that kills the woman, and the child is only an inert instrument used by the contracting uterus. In such a case the mother might be considered an aggressor materially unjust against the life of the child rather than that the child is the aggressor.

Lehmkuhl (Compendium Theologiae Moralis, 1891, p. 238) says: "Medicus graviter peccat ... si media abortus procurat: nisi quando ad salvandam matrem ex probabili opinione liceat." On page 188 he says: "Ex consulto abortum inducere, etiam liceri videtur in praesenti vitae maternae discrimine, quod per solam foetus immaturi ejectionem avert! possit ... Idque videtur applicari posse ad matrem quae tarn arcta est ut tempus praematuri partus exspectare non possit."

By foetus immaturus here he means an unviable foetus, as is evident from the context. If this probabilism of Father Lehmkuhl's stands (but it does not), a decision in most of the cases that occur in ectopic gestation would be easily made, but even he himself would not take responsibility in the matter, and that before the decision of the Holy Office which defined abortion. Since this decision, made July 24, 1895, Lehmkuhl has entirely withdrawn his opinion.

On May 4, 1898, the Holy Office published the following decree, which was approved by the Pope:

BEATISSIME PATER,—Episcopus Sinaloen. ad pedes S. V. provolutus, humiliter petit resolutionem insequentium dubiorum:

I. Eritne licita partus acceleratio quoties ex mulieris arctitudine impossibilis evaderet foetus egressio suo naturali tempore?

II. Et si mulieris arctitudo talis sit, ut neque partus prematurus possibilis censeatur, licebitne abortum provocare aut caesariam suo tempore perficere operationem?

III. Estne licita laparotomia quando agitur de pregnatione extra-uterina, seu de ectopicis conceptibus?

Feria iv, die 4 Mali, 1898.

In Congregatione habita, etc ... EE. ac RR. Patres rescribendum censuerunt:

Ad I. Partus accelerationem per se illicitam non esse, duromodo perficiatur justis de causis et eo tempore ac modis, quibus ex ordinariis contingentibus matris et foetus vitae consulatur.

Ad II. Quoad primam partem, negative, juxta decretum, Feria iv., 24 Julii, 1895, de abortus illiceitate.—Ad secundam vero quod spectat: nihil obstare quominus mulier de qua agitur caesareae operationi suo tempore subjiciatur.

Ad III. Necessitate cogente, licitam esse laparotomiam ad extra-hendos e sinu matris ectopicos conceptos, dummodo et foetus et matris vitae, quantum fieri potest, serio et opportune provideatur.

In sequenti Feria vi., die 6 ejusdem mensis et anni ... SSmus responsiones EE. ac RR. Patrum approbavit.

The third question proposed by the bishop is:

"Is laparotomy licit when performed for extrauterine pregnancy or ectopic gestation?"

The approved answer of the Holy Office to this question is:

"In a case of necessity, laparotomy for the purpose of removing an ectopic foetus (conceptus) from the abdomen of the mother is licit, provided the lives of both the foetus and the mother, as far as is possible, are carefully and fitly guarded."

The expression, "dummodo et foetus et matris vitae, quantum fieri potest, serio et opportune provideatur," is capable of various translations and interpretations.

The words might have this meaning: "In a case of necessity you may do laparotomy and remove an ectopic gestation, provided you do not kill either the mother or the foetus." If that is the interpretation, the decree means that we may never remove an unviable ectopic foetus when we know that the foetus is alive, because removal will kill it.

The sentence can also be translated in this sense: "In a case of necessity, you may do laparotomy and remove an ectopic foetus from the mother, provided you take full care to save mother and child if that is possible."

If that is the signification, it is evidently very different from the first interpretation. It would mean: do the laparotomy, remove the foetus, and if you possibly can save both mother and foetus do so, but if you can not, take the best means you can to save one or the other.

If the decree refers only to cases in which the foetus is viable, it would appear to be unnecessary—we need no decree of the Holy Office to let us do a laparotomy to remove a viable foetus. If it does not refer to a viable foetus, it refers to an unviable foetus, but to remove an unviable foetus is to either kill it or to hasten its death.

Génicot (Institutiones Theologiae Moralis, Louvain, 1902, vol. i. p. 358) has this interpretation of the decree:

"In conceptione extra-uterina licebit sane recurrere ad laparotomiam similemve operationem, quando aliqua etiam tenuissima spes affulget salvandi infantem, simul ac mater fere certo liberabitur. ... Ubi vero nulla spes hujusmodi affulget, neque in hoc casu licebit abortum directe inducere, etiamsi foetus certo moriturus sit antequam in lucem edatur, et baptismum recipere nequeat. Etenim S. Inqu., dum provocat ad responsum 19 August, 1888, satis indicat abortus inductionem a se haberi tamquam operationem directe occisivam foetus ideoque semper illicitam."

There is no question of an abortion in a laparotomy for extrauterine gestation; abortion is altogether a different operation in method and nature. Secondly, the other decree of the Holy Office to which he refers speaks of a direct killing of the foetus, but there is no direct killing of the foetus in the operation for ectopic gestation, nor is the indirect hastening of the foetus's death a means to an end. The decree on abortion is so clear it leaves no room for doubt.

Cardinal Monaco, in the Epistola ad Archiepiscopum Camarcensem, August 19, 1889, says the Holy Office decreed that "In scholis catholicis tuto doceri non posse licitam esse operationem chirurgicam quam craniotomiam appellant, sicut declaratum fuit die 28 Maii, 1884, et quamcumque chirurgicam operationem directe occisivam foetus vel matris gestantis."

Note the words "directe occisivam." Craniotomy is a direct killing, and a direct killing used as a means to an end; moreover it is an altogether unnecessary killing. Artificial abortion in the case of an unviable foetus is also a direct killing as a means to save the mother's life, but the removal of an unviable ectopic foetus is neither a direct killing, nor is it a means toward any end.

Since the meaning of the decree concerning laparotomy in extrauterine pregnancy is by no means clear, we may discuss the question until the law has been fully promulgated, ready to conform to the real meaning of the decree whenever it is explained. In that spirit we may now consider the cases that occur in ectopic gestation.

Case I. A surgeon is called in to treat a woman and he finds her in a state of collapse. He makes a diagnosis of tubal pregnancy, which has gone on to rupture with hemorrhage, and the bleeding will evidently be fatal to the mother unless it is checked. Practically the only chance of saving the mother's life is coeliotomy and the ligation of her open arteries. Dr. Howard Kelly (Operative Gynaecology, vol. ii. p. 437) says: "When the hemorrhage is sudden and excessive the patient falls in collapse; but, in spite of these alarming symptoms, she may survive a succession of similar attacks and the foetus and sac may continue to develop." This exception complicates the case slightly. If the surgeon were absolutely certain that the only possible chance to save the woman's life is coeliotomy and haemostasis, the case would be somewhat different from one in which there is some chance of escape by spontaneous haemostasis. That chance, however, is so slight, and so far beyond any means we have for forecasting, that it is mere luck, and it is to be neglected. The surgeon may safely consider the patient in the gravest actual danger.

(a) Before he opens the abdomen he can not tell whether the foetus is alive or not; but the stronger probability is that it is not, and the certainty is that it has no chance at all to remain alive more than a few minutes or hours, unless the surgeon is willing to trust to sheer luck in the expectation that he may happen to have one of Dr. Kelly's exceptions before him.

(b) The operation to save the mother is this: as quickly as possible he makes a vertical slit from four to six inches long through the woman's belly-wall. Then commonly the free blood begins to run out, or it may even spurt out some feet into the air. The surgeon can see nothing for the blood and the presence of the entrails. If the blood is not freshly welling up he bails it out with his hands or a ladle; if it is spurting he at once thrusts in his hand, feels for the foetal sac, lifts it up, and puts on clamps near the uterus on one side and near the pelvic brim on the other. This stops the hemorrhage, and he can then work more leisurely, but unfortunately this also stops the flow of blood to the foetus. He can not first examine the foetus and then stop the hemorrhage. He can not back out even if he finds a live foetus without letting the mother die on the table.

(c) If the placenta is already loose from the Fallopian tube the child is dead or it will die in a few seconds or minutes. If it was not loose the lifting out may tear it loose, and this tearing loose will hasten the death of the foetus a few minutes (but give a chance for baptising it).

(d) If the lifting out does not tear loose the supposedly fixed placenta, the foetus either will die anyhow if the mother dies, or it will die if the mother lives, because to save her the surgeon must put ligatures just where the flow of blood will be shut off from the foetus. Commonly there is no time to even look for the foetus until after the maternal arteries have been closed.

(e) The same conditions could exist in the rupture of a pregnancy in a rudimentary uterine horn as in a rupture in tubal gestation.

What is the surgeon to do in a case like this? Fathers Holaind (Amer. Eccl. Rev., January, 1894, in a note on p. 39), Lehmkuhl and Sabetti say: do coeliotomy, ligate the mother's arteries, remove and baptise the foetus.

The analysis of the case is this: (i) The action is the stopping of a fatal hemorrhage in a woman, and possibly, though not certainly, an indirect incidental hastening of a foetus's inevitable death.

(2) The object of the action is the haemostasis, which is good, and the possible indirect hastening of the foetus's death, which is evil, but, as we shall see, excusable evil.

(3) The end of the action is to save the mother's life—a good end.

(4) The circumstances are: (a) that possibly, through mere luck, the woman's condition is not necessarily hopeless: a few women have escaped in this seemingly imminent peril—but that chance of escape is not soundly probable; the stronger probability by far is on the side of a fatal issue; therefore the chance for escape may be neglected, and the woman's case may be regarded as hopeless if operation is foregone.

(b) The quickest possible work on the surgeon's part is necessary, and there is no time or chance to examine the foetus's condition before tying the maternal arteries. Before he opens the mother's abdomen he can tell nothing whatever of the foetus's condition, but the probability is all in favour of the fact that the foetus is already dead or moribund.

(c) The means are coeliotomy, and the ligation of the uterine and ovarian arteries to stop the mother's bleeding. This ligation, in the contingency that the foetus is still attached to the Fallopian tube, will also shut off the blood from the foetus, yet the uncertain shutting off of the foetal blood-supply is not intended by the surgeon as a means toward his end in any degree direct or indirect, but it is an evil circumstance associated with the action which may hasten the foetal death—even here the hastening is uncertain.

(5) The action has two effects,—one, the saving of the mother, is directly intended and evidently good; the other, the possible indirect hastening of the foetus's death, may or may not be evil. The moral centre of the whole case is this possible hastening of the foetus's death. If that possible hastening is licit the whole action is licit; if it is not permissible it

will vitiate the entire action.

Suppose that there is no doubt that the ligation of the maternal arteries in this case really hastens the foetus's death some minutes: it would still be an indirect volition. Father Lehmkuhl also calls it indirect and licit. Father Sabetti denied that it is indirect, but he held that it is licit for another reason. Sabetti said (Aner. Eccl. Rev., August, 1894): "It is evidently false to say that a means which is directly adopted for obtaining an end is only indirectly contained in the intention of the agent who so adopts it." That is true, but the minor proposition in a syllogism drawn from that statement is to be emphatically denied. The cutting off of the foetal blood is a fact associated with the means, not a means direct or indirect toward the end, which is to save the mother—the means to save the mother is the stopping of her bleeding.

This is not hair-splitting in the opprobrious sense of that term. The bases of all sins are absolutely abstract principles, and because abstract principles can not be pinched or weighed, they have often little meaning for the opposition in an argument. There is only the width of a hair between Heaven and Hell at many places along the frontier, and there is only the difference between a direct or an indirect volition separating murder and a good deed. The best ethics frequently consists in delicate hair-splitting; and despite the protests of sentimentalists, one of the most valuable benefits of Moral Science is to show us how to handle moral poisons for good purposes, as a physician uses the material poisons, opium and aconite.

If the foetus in this case of rupture in ectopic gestation were a materially unjust aggressor on the mother's life, the indirect hastening of its death would be justifiable according to all moralists, and the direct hastening would be licit according to Cardinal de Lugo, who was, in the opinion of St. Alphonsus, "post D. Thomam inter alios theologos facile princeps" (Th. Mor., lib. 4. n. 552).

Sabetti held that the foetus is a materially unjust aggressor. His reason for this opinion is that the extrauterine foetus is not in a position in which it has a right to be. If it were in the uterus, its natural position, it would have a right to its position. Ectopic gestation is a disease, not a physiological condition.

Father Aertnys (Amer. Eccl, Rev., July, 1893) denies that the foetus is an aggressor materially unjust. He says: "Nequaquam enim mortem intentat matri, sed actione, quam non ipse sed corpus matris producit, conatur ad lucem pervenire, et iste conatus non nisi ex naturali concursu rerum fit matri causa mortis. Infans ergo non est aggressor et multo minus est aggressor injustus. Hinc nego paritatem cum homine mente capto, qui delirans alteri mortem intentat; hic enim agit motus a sua voluntate, licet absque culpa, et ponit actiones in se injustas, utpote ad necandum directe intentas."

In the same periodical (January, 1894) while repeating this statement he says: "Sive in utero existat sive alibi reconditus sit [sc. foetus], nequaquam mortem intentat matri, siquidem non ipse actione propria conatur egredi, sed corpus matris infantem expellit et haec expulsio a matre emanans fit matri causa mortis."

What Father Aertnys says in these two passages is true of an intrauterine foetus, but it is altogether erroneous when applied to an extrauterine foetus, of which alone there is question here. In extrauterine pregnancy the uterus or any other part of the maternal body does not "try to expel" the foetus; the uterus has nothing at all to do with the case—the very name of the condition is extra-uterine pregnancy. If an ectopic gestation goes on to term (a very rare happening), there will be false labour and uterine contractions, and these cease after a time without effect one way or the other; but in all cases of rupture and the like the uterus is outside the question and the mother is passive. There is no attempt by the mother in extrauterine pregnancy at expulsion either before rupture or at any other time unless the dead foetus putrefies, and the maternal tissues "try to expel" it as a foreign body by breaking down into an abscess. The foetus simply grows, and its bulk bursts the tube. If it were in the uterus, the uterus would enlarge synchronously with the foetus and there would be no rupture, but the tube will not give beyond a certain point, therefore it bursts.

In normal uterine pregnancy at term the uterus and other maternal muscles are the active factors in expelling the foetus—the foetus is passive. In ectopic gestation the foetus is active, the mother is passive, and there is no attempt at expulsion from either side. In this case the foetus in the tube through the action of its own vital principle draws nourishment from the mother and grows gradually larger till it bursts the tube (it may even move its arms and legs if advanced), and this rupture tears open arteries wherethrough the mother bleeds, commonly to death. This is evidently material aggression.

Father Aertnys says the foetus differs from the murderous lunatic in this, that the madman is moved by his will, although blamelessly, in doing unjust actions directly intended as homicidal. The fact that the lunatic uses his will has no weight whatever in permitting me to defend my life against him, it is an accidental thing outside the question; but Father Aertnys in mentioning the madman's will means solely, if I understand him, that the madman is really an active aggressor. The foetus, however, is also an active aggressor without using its will. I might fall from a height toward a man and certainly endanger his life while I was not using my will at all, not conscious of the man's presence under me, or even while I was using all the power of my will against the result. In any of these cases I should be a materially unjust aggressor; and if in trying to prevent my body from killing him the man killed me, he would be blameless.

Now, in the first place, the tubal foetus is an aggressor; and since, secondly, its position is unnatural, monstrous, a disease, a thing not intended by nature, it has no right to its position, and it is therefore a materially unjust aggressor. Since it is an aggressor on the very life of the mother in a place where it should not be, the surgeon therefore may at the least stop the fatal bleeding it causes. If the foetus dies as an unwished for, though permitted, consequence of this haemostasis, the surgeon may lament this result, but he is blameless.

The foetus was blocked in its unnatural position through a defect in the mother, nevertheless it remains a materially unjust aggressor. If I by an accidental blow had made a man insane, and later this lunatic tried to kill me, I, or my legitimate protector, might lawfully kill the lunatic in defence of my life. This is an exact parallel to the case of the mother and the extrauterine foetus.

The extrauterine foetus is not like a foetus in a craniotomy case. Where there might be question of craniotomy the foetus is not an unjust aggressor even materially, as has been said: first, because it is not an aggressor in any manner, it is altogether passive; secondly, it has a perfectly natural right to be where it is. In ectopic gestation with fatal rupture the foetus is, first, an active aggressor; secondly, it has no right to be where it is. In craniotomy the foetus is killed as a direct means toward the end that its head may be reduced and extracted and the mother saved; in extrauterine gestation with fatal rupture the foetus is incidentally killed as a consequence of the haemostasis, and not as a means in any sense of the term. In craniotomy the child is wantonly killed since there are other means of saving the mother; in extrauterine pregnancy with fatal rupture the hastening of the death of the child is unfortunately associated with the only possible means we have to save the mother.

In Case I., therefore, we have an action that has an object partly good and partly, very probably, not evil; the end intended is good; the circumstances are justifiable or indifferent; consequently in Case I. the surgeon may do coeliotomy, tie the uterine and ovarian arteries, and if the foetus happens to be alive he may reluctantly and indirectly permit the hastening of its death after attempting to baptise it.

Case II. The conditions presented in Case I. are the ordinary and most common that the surgeon meets with in treating ectopic gestation, but other conditions may be found.

Suppose the surgeon, before operation, diagnoses a case of ectopic gestation, but that he can not tell whether or not the foetus is alive. The probability leans toward the side that the foetus is alive, because there is no indubitable history, as physicians say, of maternal symptoms that indicate rupture.

Medical authorities tell him to do coeliotomy at once, ligate the uterine and ovarian arteries, and remove the foetus. Would he certainly or probably be justified in following out this medical doctrine?

The mother is in actual, very probable danger of death, but not in actual, certain danger of death. She may possibly escape if operation is deferred; she has a negligible chance of escape if no operation is performed after the death of the foetus; coeliotomy and ligation of the uterine and ovarian arteries give her by far the surest chance of escape, so sure an opportunity for escape when performed early that it can scarcely be called a mere chance.

If operation is deferred the chances for rupture are about 22 per centum, say, one and a half in five chances, and all ruptures are not necessarily fatal. The chances of the mother's death, however, are much higher than that, because death can come in ectopic pregnancy from causes other than rupture. From 63.1 to 68.8 per centum (say, 66.3 per centum) of ectopic gestations treated by the expectant method result in death to the mother—just two-thirds of the women die. A. Martin in a series of 265 cases of ectopic gestation where the expectant treatment was employed found a maternal mortality of 63.1 per centum; Parry in 500 similar cases found a mortality of 67.2 per centum; and Schauta in 241 cases a mortality of 68.8 per centum.

In the 87 years between 1809 and 1896, 77 cases of coeliotomy for the delivery of viable ectopic foetuses were reported in all medical literature with a maternal mortality of about 58.3 per centum. Between 1809 and 1888 there were 37 coeliotomies with a maternal mortality of 86.5 per centum. Between 1889 and 1896 there were 40 such operations, with a maternal mortality reduced to 32.5 per centum by modern surgical methods.

The results as regards the children were almost the same in the two series, and perhaps a little better in the latter series. In the first series the 37 children were alive at delivery: the length of time in which three of these children lived is not given; three more were alive but they did not breathe; the others lived from a few seconds to days, weeks, months or years. One was well at six months, another at one year, another at seven and a half years, another in its fourteenth year, another in its fifteenth year. In the second series the results as regards the children were, as has been said, almost the same. The 40 cases that were reported from 1889 to 1896 are the standard for this phase of ectopic gestation, because they come under the diagnosis and treatment of the present day. They represent closely all such cases that occurred in the entire world between 1889 and 1896, because physicians report these operations to medical societies, and active physicians are almost without exception members of such societies— outside the civilised world these operations do not take place. In the seven years there were annually less than six cases of coeliotomy for ectopic gestation at term in the world, therefore operations at term may be neglected in discussing Case II., and the argument may be confined to the ordinary cases of expectant treatment. Schrenck in 1892 collected 610 cases of ectopic gestation which had been reported between 1887 and 1892; during the same time there were 23 cases (less than 4 per centum) of operations for the delivery of viable foetuses.

If the physician that has made the diagnosis in this Case II. leaves the patient, she may have a fatal hemorrhage at any moment. Dr. Howard Kelly reports (Operative Gynaecology, vol. ii. p. 438) a fatal hemorrhage in two days from rupture where the foetus was only as large as a Lima bean. The hemorrhage may be so suddenly fatal that the woman drops to the floor unconscious just as if she had been shot. Dr. Harris (International Cyclop. of Surgery, vol. vi. p. 784) tells of a case where three of the best obstetricians in Philadelphia met in consultation daily for 16 days expectantly watching development, but the woman died from hemorrhage in thirty minutes before any of these physicians could be called to her aid. Death may be brought about by anaemia after repeated hemorrhages. Some hemorrhages can be mistaken for colic by the physician, and this error will defer until too late the treatment for hemorrhage.

If the woman is living in a hospital where there is a resident surgeon with instruments ready, she has a better chance than if she is in her own house. Even if she has a surgeon within call the outcome of the case for her will depend largely on his skill, his presence of mind, the preparedness of his instruments, the general condition of the patient, and many other circumstances.

The instruments, ligatures, gauzes, solutions, dressing, etc., for coeliotomy are multitudinous, and all must be sterile, or the woman will be killed by septicaemia even if the hemorrhage is stopped. It is almost impossible to keep a set of instruments and the other things used in a coeliotomy always sterile and ready for instant use.

The skin surface of the patient's abdomen must be sterilised, or pus infection will get into the peritoneum through the wound. In all ordinary coeliotomies this surface is carefully sterilised by a long process the night before the operation, a protective dressing is put on, and the sterilisation is repeated the next day just before the operation. This is so important that its voluntary omission is malpractice. In the hurried operation for tubal rupture there would be no time for sterilisation of the abdominal skin surface, and probably no time to sterilise the instruments and other things used, especially the surgeon's hands.

The surgeon to do any coeliotomy needs assistant physicians—one to anaesthetise the patient, and at the least one other to work with him in the operation. He should have three or four physicians and one or two nurses. He can not do a coeliotomy alone. Hence the patient in a ruptured extrauterine pregnancy must have at the very least two physicians within call.

The woman, then, in Case II. before operation has one chance in three of life if no operation is done until the child is viable, and if she remains alive till the child is viable (when she must be operated upon) her chances for life will be no better, judging from modern statistics.

At any moment, therefore, she is in actual peril of death by two chances in three, and probably more if all special circumstances are considered. The foetus is a materially unjust aggressor in this case before rupture or other similar mishap, as it was in Case I., but not to the same extent. In Case II. it is a materially unjust aggressor as two is to three; in Case I. it is a materially unjust aggressor as three is to three.

If a lunatic is just about to fire three cartridges at me, I may know the chances are only two in three, or even only one in three, that he will hit me fatally, nevertheless I may licitly kill him to stop the firing and save my life. The mother in Case II. is in exactly similar danger of life.

The objection that the danger to my life from the action of the lunatic exists hic et nunc and that the danger to the mother's life does not threaten hic et nunc, is not of any weight. She is in actual danger hic et nunc, even while the surgeon is in the room examining her. Moreover, the matter of time here is accidental. If you give a man a poison that may kill him in ten hours, or one that may kill him in ten days, the action is essentially the same.

I am of the opinion that if this second case were proposed to moral theologians many of them would decide that the surgeon should explain the case fully to the patient or her family, and if immediate operation were insisted upon he should withdraw from the case. Nevertheless, as far as I can see, he has sound probabilism on the side that operation is justifiable.

But, it may be objected, in Case I. the surgeon ligated the uterine and ovarian arteries to stop an actual hemorrhage, and he permitted the death of the foetus; in Case II. there is no hemorrhage yet, there may possibly be none at all. I answer that in Case II. if he operates he ties the two arteries to forestall an imminent hemorrhage which might begin within the next hour if it were not securely shut off, and to forestall sepsis by leisurely and proper precautions, and exactly as in the first case he permits the death of the foetus, he indirectly kills an unjust aggressor. If the lunatic is aiming at me I do not have to wait until he begins firing to licitly shoot at him. The sooner I shoot, servato moderamine inculpatae tutelae, the more prudent my action.

To put it in another form—in Case II. the surgeon is standing before a dam (the stretched Fallopian tube) that is threatening to break at any moment and cause death to a woman below it, because there is a lunatic (the foetus) behind it tearing away the masonry. If the surgeon shunts off

32

the water just above the dam (the ligation of the arteries), he will suddenly let the lunatic who is tearing away the masonry fall down to the rocks at the bottom of the dam and be killed. May he let the lunatic fall? Certainly he may. But perhaps the lunatic will not succeed in tearing away the masonry. He is well provided with tools to do so; the chances are even two in three that he will succeed. Is he or the woman to be given the benefit of the doubt? The woman, by all means; she has a doubt worth in juridic value at the least twice as much as that which the lunatic has.

In any case of ectopic gestation the foetus has a very faint chance indeed of even living long enough for baptism if the expectant treatment is employed. We have seen that between November 1809 and November 1896 there were reported 77 cases of operation for the delivery of viable foetuses. Eleven of these children survived, 67 died within a few months, and many of these died just after delivery. Still, probably all might have been baptised. Judging, however, from the geographical distribution of the cases (see Kelly's Operative Gynaecology, vol. ii. p. 458) and the names of the operators, only about 14 of these children received baptism.

Now, since Schrenck found 610 ectopic gestations reported in five years, this indicates that the average number of cases of ectopic gestation which occur in the civilised world is at the least 122 a year, for many more (twice as many, at the lowest estimate) are not diagnosed or not reported when diagnosed. In the 80 years, then, between 1809 and 1896 there were at the least 9760 cases of ectopic gestation in the civilised world; in the uncivilised countries there were certainly as many more with not a child saved, or even brought out of the pelvic cavity. To be sure, by rejecting perhaps a third of the cases through bad diagnoses and neglect of reports, there were 20,000 cases; and in all these hardly 20 children baptised—one in a thousand.

Modern surgical methods and improved diagnosis will do little to better the condition, from the nature of the disease. Between 1893 and 1896 there were 21 cases of operation for the delivery of viable foetuses reported, and this list is approximately correct, because the surgeons that operate on such material are men that as a rule report their work even when it is to their discredit. In these 21 cases, 6 mothers, 28 per centum died, 72 per centum recovered. Even if modern surgery should save all the mothers who had escaped until the foetus was viable, and should bring all the children to baptism, there would not be more than about 7 such cases in the world annually. Increased skill in diagnosis would raise the number of children brought to baptism, but it would more than proportionately raise the whole number of ectopic gestations discovered. If 10 foetuses were brought from the pelvic cavity alive in the 130 cases of ectopic gestation of the year, the chances for an extrauterine foetus to only reach baptism at a viable age (not to live after baptism) are only 7 in 100 at a most liberal estimate. Statistics are unreliable, of course, but I am giving odds of two to one. The foetus has a much better chance for baptism if the coeliotomy is done as early in the pregnancy as possible, but it has a negligible chance of life in any case. Since the creation of man there have been less than 15 extrauterine children saved, and of these 15 four were less than a year old when reported, and three under five years of age: the oldest was fifteen years of age, and all were weaklings.

The practical rule, then, is that the ectopic foetus will die anyhow, and operation only indirectly (mark the word) accelerates the inevitable death of a materially unjust aggressor, while it gives the mother the best chance for her life, which is in very grave peril.

Case III. The surgeon before operation diagnoses with the help of consultors extrauterine pregnancy, but he or they can not tell whether the foetus is alive or not. What should he do?

In my opinion he may operate with much more solid probability than that which exists in Case II. If the argument is more for the death of the foetus than for its life, this, of course, strengthens the permissibility of the operation.

(1) The danger to the mother is exactly the same, caeteris paribus, as in Case II.; (2) the foetus is only probably alive. An actual danger to life is opposed to the probable life of a materially unjust aggressor; therefore the surgeon may probably operate at once. Probable here is used in the technical sense of the term.

Case IV. The following case is given because a similar one was proposed in the articles in the American Ecclesiastical Review, but it is not a practical case.

The surgeon, after consultation, does not know whether the growth in a woman's pelvis is a malignant tumour or a sac containing an extrauterine foetus. If the growth is a malignant tumour, the woman is in actual and certain danger of life, her death is a mere matter of time if a malignant tumour is not removed, and the sooner the tumour is removed the better. If operation is deferred, metastases of the tumour will have occurred, and operation will be too late. The indication when we find a malignant tumour is, if it is not already too late to operate, to take it out at once.

If the surgeon thinks that the growth may possibly be a foetus, and he puts off the operation until a time when certain signs of pregnancy should be present to establish a diagnosis of gestation, or their lack to establish a diagnosis of tumour, it would almost surely be too late to operate in the event the growth turned out to be a malignant tumour.

As has been said, the case is not practical, because malignant tumours of the tube are so very rare that they are not to be looked for,—only one or two have been observed. Malignant tumours about the tube should be diagnosed. Supposing, however, the case to stand, it offers in favour of operation a probabilism stronger than that in any case except Case I., because the mother's danger is graver, and the argument concerning the foetus is the same as that in Case III.

Case V. Suppose a doubtful case like Case III. or Case IV., but after the surgeon has opened the abdomen he finds a foetus evidently alive. This is an improbable but a possible case. Case V. then becomes like Case II. with the addition of another grave danger to the lives of both the mother and the foetus, which is the coeliotomy already performed. The suggestion that the surgeon can leave the woman, back out of the case, is absurd. If he closes the abdomen, the coeliotomy may cause tubal abortion, the wound might have to be opened again in a few hours or a few days, and the mother would be left in much greater peril than she was in Case II. For the reasons already given, he should go on with the operation.

Case VI. Suppose a case like Case V. in every particular except that when the surgeon finds the foetus he can not tell whether it is alive or not. He should, a fortiori, finish the operation.

Case VII. A case of ectopic gestation is diagnosed, the conditions are explained to the woman, and she refuses to be operated upon. Is she justified? The probability is one to two that she will escape death if she waits, and much less than one to two if she finally refuses operation. The moralists would tell her she may refuse operation.

Case VIII. Let us suppose a case where a Fallopian tube either has its lumen so narrowed by a gonorrhoeal inflammation that although the spermatozoa may pass through and fecundate the ovum this fecundated ovum can not get out to the uterus; or, secondly, that the gonorrhoeal infection has completely shut the tube, yet migratory fecundation has occurred through the route of the other tube and the passage along the fundus of the uterus to the ovary of the infected side. In either case an ectopic gestation begins.

The first case is improbable from a medical point of view, and the second is barely possible. Gonorrhoeal infection of the tubes is common enough, but when it occurs it usually shuts the tube up permanently. In chronic salpingitis at times the ovarian end of the tube is not wholly closed at once, and since the body of the ovary is very rarely affected by gonorrhoea, there is a possibility worth considering of a tubal pregnancy through migration to occur.

In such a condition the woman might have been infected with gonorrhoea, first, before her marriage through fornication or accident; second, after her marriage through adultery or accident; third, after the marriage by her husband.

If she had been infected through fornication or adultery, she is accountable for the foreseen consequences of her sin, and she has put an impediment for which she is responsible before the embryo. Suppose the physician knows these facts. Then the excuse for indirectly hastening the death of the foetus does not, at first sight, seem to exist, because the foetus is apparently not a materially unjust aggressor. It could easily happen that a surgeon's refusal to operate in a case like this would cause the death of the mother and foetus. Should he let both perish? Is he to let the mother die for the sake of staving off for a half-hour the certain death of a useless embryo the size of a pigeon's egg? It is not a useless embryo the size of a pigeon's egg, but a human being, the most important thing on earth, and a human being shut off from life and baptism as a direct consequence of that woman's brutal sensuality. But the woman may be the mother of other helpless children. What is to be done? Let us recur to the example of the homicidal maniac.

If I accidently by a blow make a man insane and that insane man afterward tries to kill me, I or my protector may permit his death to save my life. If I maliciously make a man insane and he afterward tries to kill me, may I or my protector kill him in my defence? Some may say that I may not because I have lost all juridic superiority over the madman as a consequence of my sin against him. That position, however, does not seem to be correct.

If it is correct, parity makes the assertion true that the foetus in the case supposed above may not be indirectly killed to save the mother. If it is not true, the foetus may be indirectly destroyed. Does my sin against the insane man give him a right to kill me? By no means. Nothing but defence of life or its equivalent gives any private individual the right to kill another. The man might kill me before this aggression of mine, in defence of his sanity, but after the fact such a killing would be mere revenge, or an actus hominis, not a right.

The woman, we suppose, has maliciously put the foetus in its position of material aggressor, but has the foetus the right to kill her? No; the foetus is an individual not acting in self-defence, it is merely growing. Has the woman or the surgeon, her protector, the right to permit the death of the foetus to defend the woman's life? I think they have, because the foetus here also is, from its unnatural position, a materially unjust aggressor.

But, you say, this is a vicious circle. You justify the permitted death of the foetus in Case I. because it is a materially unjust aggressor, and it is a materially unjust aggressor because it is in an unnatural position where it has no right to be; but in the present case the mother put it in the unnatural position, and it therefore has a right to be where it is. No: the consequence does not follow. The fact that the mother put the foetus in its unnatural position does not give the foetus a right to be in that position, although it constitutes a ground for her punishment by proper authority. You object again, if this woman has a right to permit the death of the foetus to save her own life, how may she be punished for that death? She will not be punished for the actual coeliotomy which indirectly caused the death of the foetus, but she will be punished for the sin of putting that child in a position in which it had to be killed. This seems to be a distinction without a difference. As far as the mother is concerned, transeat; but it is a real distinction as far as the surgeon is concerned.

If the woman's condition is a result of accidental infection before or after marriage, the case goes into the class of those discussed above, and operation is justifiable.

If her infection comes after her marriage adulterously, her sin is the greater, but the operation is justifiable for the reasons which were given in the case of culpable infection before marriage.

If she had been infected by her husband, the operation is justifiable—the father is accountable for the foetus's death.

Fortunately the entire case is so nearly hypothetical that it is little more than mere words.

AUSTIN ÓMALLEY.

PELVIC TUMOURS IN PREGNANCY

Tumours of the uterus and its adnexa at times, though rarely, complicate pregnancy, and they may involve certain moral questions that have been little discussed. The tumours that cause difficulty are ovarian and uterine.

Cystic ovarian tumours commonly do not prevent impregnation, if there has been an absence of inflammation. When these cysts are small they may not disturb pregnancy or delivery; large cysts can, however, become a source of danger. They may sink into the pelvis and block the channel of delivery needed by the child at term; they may have their pedicles twisted, and thus become gangrenous and septic. Big cysts of the ovary may during the growth of the pregnant uterus press upon the portal vein, or the diaphragm, or they may burst or cause sepsis. Litzman, in 56 cases of ovarian tumours complicating pregnancy, had only 10 normal deliveries; and Remy held that 23 per centum of these cases, when left untouched, result in death to the mothers. Stratz says the mortality is 32 per centum, and it has gone as high as 40 per centum. Some physicians teach that any ovarian cyst found complicating pregnancy should be removed surgically. Other authorities hold that they should all be treated expectantly: if they threaten the life of the mother, they should be tapped by a trocar through the belly-wall or the vagina, and removed only after labour. This second operation is safe, and I think it should prevail.

Such cysts have often been removed during pregnancy. Orgler reported 146 ovariotomies (removal of the ovaries) performed during gestation with only four maternal deaths—2.7 per centum. If the operation had not been performed about 32 per centum of these women would have died. The chance against saving the child in such an operation is the crux. If there is no operation 17 per centum of the cases result in abortion and the loss of the child, as Remy found from a consideration of 321 cases. In Orgler's series of 146 ovariotomies, where he lost only 2.7 per centum of the mothers, and saved about 30 per centum that would have died (97 per centum in all); he lost 32 children through abortion caused by the ovariotomies, or 22.5 per centum; whereas by the expectant method (without tapping) only 17 per centum of the children were lost.

Bovee of Washington, however, reported 38 cases of removal of the ovaries during pregnancy with one maternal death and only four abortions, or 12.6 per centum. That is considerably less than the loss by the expectant method without tapping. As Bovee succeeded, other men now do, but it would be far better to attempt tapping first. The earlier in the pregnancy either tapping or removal is done the better.

Fibroid tumours of the uterus, complicating pregnancy, occur in about 0.6 per centum of pregnancies, and they usually go on without causing trouble; but again these tumours may block the pelvic outlet, they may dangerously press upon abdominal viscera and the diaphragm; some writers hold they may become inflamed and degenerate with sloughing and gangrene, and thus bring about sepsis and death to the mother and child. That they become gangrenous must very rarely happen; the increased blood supply should prevent gangrene, but cause an increase in the size of the fibroma.

A group of gynaecologists maintain that when fibromata cause dangerous symptoms in pregnancy the uterus should be taken out in part or wholly if the tumour is so deeply involved in the uterine wall that it can not be separated. This operation, of course, kills the foetus. At times the child is viable, and a precedent caesarean section will save it. Surgeons do not remove fibromata merely as a precaution, as they sometimes do in the case of ovarian cysts. Other surgeons say it is safe to wait. If the channel of delivery is blocked, these men wait till term and then do caesarean section; in other cases the tumour will often be lifted up out of the way during the later stages of gestation or labour.

In those very rare cases where it is necessary to remove the uterus wholly or in part before the child is viable, and thereby also to kill the foetus, the operation at first glance seems in no wise to differ in nature from a craniotomy upon a living child. The condition, however, is commonly worse than one in which a craniotomy is indicated, because in the latter condition we have a viable child, and the caesarean section to solve the difficulty, but in the former we have a child not viable, and therefore the caesarean section would be useless, except for the opportunity it might give for baptism of the child. In such a case must the surgeon let the mother die lest he hasten the death of a non-viable child?

The action reduces to this, that the surgeon by operating would permit a hastening of the inevitable death of the foetus while saving the mother's life, but the child is not an unjust aggressor, not even a materially unjust aggressor. It has a right to be where it is. The only excuse for hastening its death is to save the mother's life,—there is no question of self-defence; but deliberately to hasten the death of a human being a second of time, except it be done by an individual in self-defence against an unjust aggressor, or by the state for legitimate cause, is murder. It seems probable, however, that there is something to be said in favour of the unavoidable hysterectomy (removal of the womb) in a pregnancy complicated with uterine fibromata that undoubtedly endanger life.

Such cases differ from craniotomy, or the direct killing of a foetus (which were formally forbidden by the Holy Office on May 28, 1884, and August 19, 1888, and always forbidden by the natural law) in several factors: first, in craniotomy the child is directly killed, although it is not an aggressor, in the hysterectomy it is permitted to die, it is indirectly killed; secondly, in craniotomy there is a viable child, in the hysterectomy, an unviable child; thirdly, in craniotomy there is a killing that is a means toward the end of saving the mother's life, in the hysterectomy there is a permitted hastening of the foetus's death, and this is only a circumstance inseparably joined to the act; fourthly, in craniotomy the killing is utterly uncalled for, because the caesarean section, or symphyseotomy (a temporary dividing of the pubic joint to get more room) will do instead, in the hysterectomy, because the child is not viable, there is no alternate way out of the difficulty; fifthly, formal judgment has been pronounced by the Holy Office in craniotomy, no formal judgment has been made as regards this hysterectomy.

Suppose A and B are on a boat hoisting a weighty object to a ship; the tackle breaks, the falling weight mortally hurts B, and wedges him fast to the wrecked boat. The boat is about to sink and drown both men, but if A tips off the weight, and with it unavoidably the entangled B, A can float to safety. A will indirectly hasten the inevitable death of B by throwing off the weight which will drag him down. May A do so? Very probably he may.

Two swimmers, A and B, are trying to save C, who dies in the water, and as he dies he grips A and B so tightly they can not shake the corpse off. A is weak, and he will soon sink and drown owing to the weight of the corpse; B also will later go down with A and C. A, however, cuts his clothing loose from the grip of the corpse (or some one in a boat does so who can do no more) and A is saved; but thus immediately B is drowned, owing to the fact that the full weight of the corpse is upon him. Is A, or the man in the boat, justified? Probably they are. A is the mother, B the foetus, C the diseased uterus, the man in the boat is the surgeon. The mother has herself cut away from the uterus and the foetus's death is hastened.

Again, take an example used by Father Ricaby in his Moral Philosophy, p. 205 (London, 1901). He supposes a visitor to a quarry to be standing on a ledge of rock which a quarryman had occasion to blast, and the quarry man saw that "unless that piece of rock where the visitor stood were blown up instantly, a catastrophe would happen elsewhere, which would be the death of many men, and if there were no time to warn the visitor to clear off who could blame him if he applied the explosive? The means of averting the catastrophe would be, not that visitor's death, but the blowing up of the rock. The presence or absence of the visitor, his death or escape, is all one to the end intended: it has no bearing thereon at all."

If these examples of indirect killing are allowable, why may not the surgeon in the rare example presented here remove the uterus and indirectly permit the hastening of the foetus's death? That hastening of death is not an end, nor a means toward an end, but a circumstance only reluctantly and indirectly willed. The end is to save the mother's life, and the means is the removal of a septic or impacted uterus.

It may be objected that an artificial abortion wherein the womb is emptied of an unviable foetus to save the mother's life is only an indirect hastening of this foetus's death, but there is a difference: in abortion the removal of the foetus is the means whereby the end is attained, in the hysterectomy the removal of the tumour is the means whereby the end is attained. This argument is advanced only tentatively and with diffidence, that the matter may be discussed and settled by authority.

Sometimes carcinoma (a cancer) complicates pregnancy—once in 2000 cases is above the average. A carcinoma is a malignant tumour, and the malignancy is made much worse by the stimulus of pregnancy with its increased blood supply. The maternal deaths from carcinoma of the uterus during pregnancy is, according to the latest and most favourable statistics, 30 per centum. The mortality of the children is from 50 to 63 per centum.

Now, first, if an artificial abortion is induced while the foetus is unviable, the foetus is lost and the mother's condition is not materially improved.

Secondly, if curettement (a scraping away with a sharp spoonlike instrument), cauterization, or amputation of the uterine cervix are performed, the mother is helped very little, if at all, and consequent abortion is frequent.

Thirdly, if caesarean section is done at term the child has a good chance (Sanger saved 16 of 18 children thus in one series: over 88 per centum), but this operation nearly always kills the mother when cancer is present, unless the entire uterus can be removed, and often it can not be removed; that is, the case is inoperable and removal is useless owing to extension of the cancer into the surrounding tissues.

Fourthly, if the mother's condition is hopeless, a caesarean section gives the child a chance for life, but the operation will hasten the mother's death in nearly every case.

The first and second cases here are not practical. If the surgeon can remove the uterus at term after a caesarean section, that is the most reasonable operation for the mother and child, and it offers no moral difficulty.

If the mother's condition is so bad that the uterus may not be removed, the chances are that her death will be hastened by caesarean section, but if caesarean section is not done, from 50 to 63 per centum is the ratio against the saving of the child. I do not think a general rule can be given as regards the certainty of hastening the maternal death: the reckoning is to be made to meet the particular condition. It seems, however, probable that in every case of inoperable carcinoma of the uterus complicating pregnancy a caesarean section would hasten the maternal death. She will die anyhow from the cancer, but in certain cases she may live longer if the section is not done.

If, again, a carcinoma of the uterus is inoperable at term, the delivery of the child may be impossible without caesarean section, from uterine inertia, or the opposition of the dense inflamed tissues, or the friability of these tissues. In such a case without the section she would die, and die probably sooner than with it. The operation would possibly slightly prolong her life, by, say, a few hours or days, and it certainly would give the child a very good chance for its life. She may, of course, die upon the operating table, but she would die in childbed without the section.

The case is different from the ordinary caesarean section done because of a narrow pelvic bony girdle. In the latter condition the chances that the mother will live are very high if the surgeon is competent, but in the carcinoma case she will die no matter who the surgeon may be, and very probably, or almost certainly, her death will be hastened by the operation in the majority of cases.

If the condition is such that the woman can not be delivered without the section, I see no difficulty against operation, because the surgeon can not, as far as I know, say positively whether he will hasten the maternal death or not, and in the circumstances he may take advantage of the doubt.

If the woman with an inoperable carcinoma uteri may be delivered without section, should such a delivery be chosen although it raises the chances of mortality as regards the child from about 12 per centum to at the least 50 per centum? It is a matter of a very probable hastening of the mother's death as weighed against the safety of the child—the child has about one chance in two of life without the section, and, say, seven chances in eight with the section. The operation is far preferable as regards the child alone, but not preferable as regards the mother alone. Is it then allowable?

In the hysterectomy for fibroma already considered, the mother is saved and the child's inevitable death is certainly hastened; in the caesarean section the child is most probably saved, and the mother's inevitable death is most probably hastened; we might say, in some cases, that her death is undoubtedly hastened. If in the carcinoma case here the child had no chance whatever for delivery except by the caesarean section, while the mother's death would be probably or certainly hastened, she might legitimately consent to the operation or she might legitimately refuse the operation.

The child, however, has, as we said, one chance of delivery in two without the section, while the mother's death will very probably be hastened. If the mother's death would certainly be hastened by the section, her death, although it would be a circumstance and indirect, not an end nor a means, would not have counterbalanced against it necessarily the saving of the child's life, because the child has one chance in two in any event. In such an hypothesis the operation seems to be unjustifiable.

If, however, the hastening of the mother's death is only probable and not certain, may we oppose that probability to the advantage that must accrue to the child through the section? If the doubt that her death will be hastened is soundly probable, the woman may consent to the operation. She risks through charity the hastening of her own death for a great advantage to the child, but she may risk legitimately immediate death in major surgical operations for an advantage less than the saving of life itself. She may have her skull opened for the removal of a depressed bone that is causing paralysis, she may have her knee-joint opened for the wiring of a patella to prevent lameness, but both these operations always immediately endanger life. She may go into a burning house, jump into a river, and so on, to save her child from possible injury.

AUSTIN ÓMALLEY.

ABORTION

If pregnancy ends in the emptying of the uterus before the sixteenth week of gestation, the condition is called an abortion; if this happens between the sixteenth and the twenty-eighth weeks, it is miscarriage; if the child is born after the twenty-eighth week but before full term, the birth is premature. The term "abortion" in the popular mind carries with it the notion of criminal interference, and the word "miscarriage" is used for both abortion and miscarriage by the laity; physicians, on the other hand, commonly use the term "abortion" for both abortion and miscarriage. These conditions may occur spontaneously or they may be induced artificially.

Spontaneous abortions are very frequent; perhaps one in every five or six pregnancies is the proportion: the writer has known a single physician, not a specialist in obstetrics, to be called to three in one day and that in private practice. From 150 to 200 children in every 1000 that are conceived never get a chance for baptism. In the early months of pregnancy the foetus is usually dead before expulsion takes place. Twisting of the cord, hydramnios, syphilis, an acute infectious disease in the mother, poisonings of the mother by metals and the like substances, maternal cardiac and renal diseases, chronic inflammations and displacements of the womb, and violent emotions are some of the causes of abortion. In certain women a slight exertion, a misstep, a fall, a ride over a rough road, the debitum conjugale, and similar causes bring on abortion; in other women almost no shock is enough to make them miscarry. Inflammations and displacements of the womb cause most of the abortions in the first four months, and after that time syphilis and Bright's disease are the chief forces at work.

If a woman in early pregnancy begins to lose blood from the uterus, and has pain in her back and lower abdomen, abortion is threatened; if this hemorrhage is marked, and the cervix is dilated, the abortion will very probably occur; and the escape of the liquor amnii renders the abortion unavoidable. In this latter case the vagina and the cervical canal are packed with sterile gauze to check the hemorrhage, and after twenty-four hours it is removed. Then commonly the entire ovum comes away with the gauze, or what remains of it is taken out with a curette.

Valvular lesions of the heart in pregnancy make a maternal mortality of about 28 per centum, according to Guérard, and when compensation is lost the mortality may run from 48 to even 100 per centum with different physicians and different cases. The prognosis is good as long as compensation is retained, but very bad if this fails. In the latter condition premature labour is indicated, or the early removal of the viable child. Catholic physicians may not induce artificial abortion of an unviable foetus. The decree of the Holy Office concerning this matter is as follows:

Beatissime Pater,—Stephanus … Archiepiscopus Cameracensis … Quae sequuntur humiliter exponit:

Titus medicus, cum ad praegnantem graviter decumbentem vocabatur, passim animadvertebat lethalis morbi causam aliam non subesse praeter ipsam praegnationem, hoc est, foetus in utero praesentia, una igitur, ut matrem a certa atque imminenti morte salvaret, praesto ipsi erat via, procurandi scilicet abortum seu foetus et ejectionem. Viam hanc consueto ipse inibat, adhibitis tamen mediis et operationibus, per se atque immediate non quidem ad id tendentibus, ut in materno sinu foetum occiderent, sed solummodo ut vivus, si fieri posset, ad lucem ederetur, quamvis proxime moriturus, utpote qui immaturus omnino adhuc esset.

Jamvero lectis quae die 19 Augusti, 1888, Sancta Sedes ad Cameracenses Archiepiscopos rescripsit: tuto doceri non posse licitam esse quamcumque operationem directe occisivam foetus, etiam si hoc necessarium foret ad matrem salvandam: dubiis haeret Titius circa liceitatem operationum chirurgicarum, quibus non raro ipse abortum hucusque procurabat, ut praegnantes graviter aegrotantes salvaret.

Quare ut conscientiae suae consulat supplex Titius petit: utrum enuntiatas operationes in repetitis dictis circumstantiis instaurare tuto possit.

Feria iv, die 24 Julii, 1895.

In Congregatione generali S. Romanae et Universalis Inquisitionis … Emi ac Rmi Domini Cardinales … respondendum decreverunt: Negative, juxta alias decreta, diei scilicet 28 Maii, 1884, et 19 Augusti, 1888.

… Sanctissimus Dominus noster … approbavit.

Other documents referring to the same matter are the following:

Epistola ad Archiepiscopum Cameracensem. … Anno 1886, Amplitudinis tuae Praedecessor dubia nonnulla hinc supremae Congregationi proposuit circa liceitatem quarumdem operationum chirurgicarum craniotomiae affinium. Quibus sedulo perpensis,

Eminentissimi ac Reverendissimi Patres Cardinales una mecum Inquisitores Generales, feria iv, die 14 currentis mensis, respondendum mandaverunt:

In scholis catholicis tuto doceri non posse licitam esse operationem chirurgicam quam craniotomiam appellant, sicut declaratum fuit die 28 Maii, 1884, et quamcumque chirurgicam operationem directe occisivam foetus vel matris gestantis.

Idque notum facio Amplitudini tuae, ut significes professoribus facultatis medicae Universitatis catholicae Insulensis. ...

Romae, die 19 Augusti, 1889. ...

R. CARD. MONACO.

The date of this response here is 1889, but in the preceding decree it is given as 1888. In the Acta Sanctae Sedis the date is 1889.

Another letter from Cardinal Monaco is this:

Eme et Rme Dne,—Emi PP. mecum Inquisitores generales in Congregatione habita feria iv, die 28 labentis Maii, ad examen revocarunt dubium ab Eminentia tua propositum—An tuto doceri possit in scholis catholicis licitam esse operationem chirurgicam, quam Craniotomiam appellant, quando scilicet, eâ omissâ, mater et infans perituri sint, eâ e contra admissâ, salvanda sit mater, infante pereunte?

Ac omnibus diu et mature perpensis, habita quoque ratione eorum quae hac in re a peritis catholicis viris conscripta ac ab Eminentia tua hinc Congregationi transmissa sunt, respondendum esse duxerunt: Tuto doceri non posse.

Quam responsionem cum SSmus D. N. in audientia ejusdem feriae ac diei plene confirmaverit, Eminentiae tuae communico. ...

R. CARD. MONACO.

Romae, 31 Mail, 1884.

Emo Archiepiscopo Lugdunensi.

Another decree concerning abortion is in part as follows:

Beatissime Pater,—Episcopus Sinaloen. ad pedes S.V. provolutus, humiliter petit resolutionem insequentium dubiorum:

I. Eritne licita partus acceleratio quoties ex mulieris arctitudine impossibilis evaderet foetus egressio suo naturali tempore?

II. Et si mulieris arctitudo talis sit, ut neque partus praematurus possibilis censeatur, licibitne abortum provocare aut caesaream suo tempore perficere operationem? ...

Feria iv, die 4 Mail, 1898.

In Congregatione habita, etc. ... EE. ac RR. Patres rescribendum censuerunt:

Ad I. Partus accelerationem per se illicitam non esse, dummodo perficiatur justis de causis et eo tempore ac modis, quibus ex ordinariis contingentibus matris et foetus vitae consulatur.

Ad II. Quoad primam partem, negative, juxta decretum Feria iv, 24 Julii,

1895, de abortus illiceitate. Ad secundum vero quod spectat; nihil obstare quominus mulier de qua agitur caesareae operationi suo tempore subjiciatur. ...

In sequenti Feria vi, die 6 ejusdem mensis et anni ... SSmus responsiones EE. ac RR. Patrum approbavit.

Pyelonephritis (an inflammation of the kidney where pus is present), from the pressure of the pregnant uterus, is a condition which sometimes obliges the physician to bring about premature labour to save the mother. The symptoms usually appear in the latter half of gestation.

Chorea ("St. Vitus' Dance"), when it develops during pregnancy, has a maternal mortality of from 17 to 22 per centum. It may cause death before the child is viable, and to empty the uterus will stop the symptoms. Here the decrees of the Holy Office will occasionally prevent the Catholic physician from interfering.

If a grave surgical operation is imperatively indicated during pregnancy, and may not be put off until after delivery, it should be undertaken in many cases, because modern technique commonly does not bring about an abortion; but, in general, no rule can be given—each case must be judged separately.

If a pregnant woman has at the same time considerable albumen in her urine and a low excretion of urea, her condition is very dangerous. To empty her uterus will, in most cases, relieve the renal trouble, but in any case premature labour is not to be induced rashly: many women escape, when by all the rules they should die.

Eclampsia is a very grave complication of pregnancy, and it was formerly supposed to be uraemia. The disease is characterized by convulsions, loss of consciousness, and coma. It occurs, commonly, in the second half of gestation, but it has been observed as early as the third month. About 70 to 80 per centum of the cases are in primiparous women. The convulsions may come on altogether unexpectedly, but commonly the attack begins with symptoms of toxaemia. Eclampsia may occur before, during, or after parturition. When it comes before term it usually ends in spontaneous or artificial abortion, but at times the woman dies undelivered. Now and then she may recover and be delivered at term.

The kidneys are usually affected, even in those cases in which albuminous urine is not found. There is also a hemorrhagic inflammation of the liver; and oedema and congestion of the brain, with or without apoplexy, are other symptoms of the disease. There are other lesions, but the chief are in the kidneys, liver, and brain.

The aetiology of the disease is not yet known, and there are very many theories offered to explain it. The prognosis is always serious, and the condition is one of the most dangerous found in pregnancy. The mortality varies, but it is about from 20 to 25 per centum in the women, and from 33 to 50 per centum in the children. It is impossible to determine the prognosis in particular cases, but a large number of quickly recurring convulsive seizures, with a weak, thready pulse, and a high temperature usually indicate a fatal ending. Apoplexy, oedema of the lungs, and paralysis also, as a rule, end in death.

If the uterus is emptied during the convulsions, these cease either immediately or soon after delivery, in from 66 to 93 per centum of the cases, and the maternal mortality then is about 11 per centum. With the expectant treatment, in convulsive cases, about 28 per centum of the women die, although a use of aconite in these cases may better the prognosis.

Pernicious vomiting (hyperemesis gravidarum) is another complication of pregnancy, which sometimes results fatally if the uterus is not emptied. There are cases, especially those with high fever, which end in death despite all treatment. Here, again, the aetiology of the disease is not known. There is commonly an element of hysteria in the condition, and in such a case moral suggestion often has a curative effect Any bodily irritation is to be removed. Eye-strain alone is enough to cause persistent vomiting. It is very difficult to decide when premature labour is absolutely indicated, because some very bad cases recover spontaneously when all hope is lost.

Hydramnios, or an excessive quantity of liquor amnii, may so distend the uterus as to cause grave danger to maternal life, and if the child is viable the uterus should be emptied.

Intrauterine hemorrhage brought on by a premature separation of the placenta is a very dangerous condition: 32 to 50 per centum of the mothers die, and 85 to 94 per centum of the children. In a marked hemorrhage the only way to save the mother is to empty the uterus, so that it may contract and thus close the patulous vessels.

Placenta praevia is a placenta implanted in the neighbourhood of the internal os of the uterine neck. This is a very perilous condition, calling for the induction of premature labour. The medical treatment is artificial abortion as soon as the condition is diagnosed in any stage of gestation; but this is, of course, in conflict with the decrees of the Holy Office. Under expectant treatment about 40 per centum of the mothers die, and 66 per centum of the children. Those children that are born alive commonly die within ten days after delivery. The great foetal mortality is due to premature birth and asphyxiation. Skilful obstetricians get much better results, but skilful obstetricians are unfortunately rare.

When the grave complications enumerated above occur in the early months of pregnancy, before the foetus is viable, the Catholic physician, since by the natural law and the decisions of the Holy Office he is forbidden to induce artificial abortion, must withdraw from the case. If there is no other physician to attend to the woman, he must let her die. He can not withdraw without explanation, and in many cases the explanation of the condition will promptly result in the calling in of a physician who has no scruple in inducing this abortion, no matter how reputable he may be. The universal medical doctrine is to induce abortion in cases where abortion will save the mother's life and the foetus is "too young to amount to anything." This is looked upon as legitimate abortion by the very best men that do not recognise the authority of the Holy Office: they deem the position of the Catholic physician in these cases as altogether erroneous, or even criminal.

The position of the Catholic moralists on craniotomy has turned the attention of many non-Catholic physicians to the immorality of the act, which formerly was deemed entirely permissible. Probably the same good result will be effected in the matter of abortion.

AUSTIN ÓMALLEY.

THE CAESAREAN SECTION AND CRANIOTOMY

In the caesarean section the infant is delivered through an incision in the abdominal or uterine walls. The operation, according to one opinion, takes its name from Caius Julius Caesar, who, it is said, was brought into the world in this manner, "a caeso matris utero"; this, however, is a myth.

Up to 1876 the maternal mortality from the operation was about 52 per centum. Between 1787 and 1876 in the city of Paris there was not one successful caesarean section as far as the mothers were concerned. At present on an average less than 10 per centum of the women are lost, and expert surgeons have better results. Up to about 1902 Zweifel had made 76 such sections with only one death, and Reynolds, 23 with no death. Leopold has performed the operation four times on the same woman, and Ahlfeld and Birnbaum have reported instances where the same woman has had five caesarean sections performed upon her. The operation is, of course, capital, and always most serious, even in city hospitals.

The indication for the operation is chiefly a narrow pelvis, which blocks the delivery of the child. There are no reliable statistics as to the frequency of narrow pelves in the United States; but Dr. Williams, of the Johns Hopkins University Hospital, in a series of 2133 cases found 6.9 per centum in white women and 18.82 in negroes. Normally the average female pelvis, at its narrowest diameter, is 11 centimetres wide. This part is called the conjugata vera, and it is the diameter from the promontory of the sacrum behind to a point on the inner surface of the symphysis pubis in front.

In delivery much depends upon the size of the child, and in each case the obstetrician waits until he sees that delivery is impossible by natural means before he resorts to the caesarean section or other operative interference. Of two women with pelves of the same contraction one may require the section and the other may have a normal labour. A bisischial diameter at the outlet of the parturient canal of 7 centimetres or less is an indication for section; so are certain tumours that block the delivery of the child.

When the conjugata vera is less than 7 centimetres in flat pelves, or 7.5 centimetres in generally contracted pelves, the treatment varies in the customary medical practice according as the child is alive or dead, and it varies as the condition of the mother. The common medical doctrine will first be given here before the moral questions that may be involved are mentioned.

If the deformity is diagnosed during pregnancy, the woman is sent to a hospital, the caesarean section is performed, and thus all the children, and nearly all the mothers, are saved. When the narrowness of the pelvis is discovered only during labour, the treatment varies with the condition. If the woman is not septic, and has not been repeatedly examined by the vagina, and if the surroundings are favourable, caesarean section is done; if she is septic, the indications are for the section, or symphyseotomy or craniotomy. Where the conjugata vera is below 5 centimetres in length, the caesarean section is the only method to get the child out, dead or alive, and after the child has been delivered, the uterus, if septic, is removed. If the conjugata vera is at the least 7 centimetres long, symphyseotomy may be done; if the conjugata vera is above 5 centimetres, the mother septic, and the child dead or dying, craniotomy is indicated. Even if the child is not dying, some obstetricians will do craniotomy.

In cases where the conjugata vera is above 7 centimetres in flat pelves and 7.5 centimetres in generally contracted pelves, the treatment can not be reduced to general rules. Delivery without operation occurs in many of these cases, but commonly the condition is obscure to the physician for some time. We can measure the pelves, but the size of the child's head is not satisfactorily measurable.

If the conjugata vera is from 10 to 9 centimetres, or from 9.5 to 8.5 centimetres, labour without operation is the rule, and the child can usually be delivered by forceps. Should the child die during labour in these cases, it is best delivered by craniotomy, unless the longer diameter of its head has already passed the narrowest part of the pelvis.

When the conjugata vera is from 8.9 to 7.5 centimetres, about 50 per centum of the women will be delivered with forceps, but the other half will not. After about two hours of the second stage of labour delivery by forceps is tried, but prolonged traction is not applied. Occasionally delivery will come when least expected, but often it will not. If the head sticks, caesarean section is done in favourable circumstances, and craniotomy in unfavourable circumstances. If there is ground for supposing that septic infection of the mother has begun, the conditions are explained, and if she wishes to have the caesarean section done the risk is left to her. When the breech or face of the child presents in contracted pelves, the condition is especially unfavourable for the child.

There are very many varieties of deformed pelves, but the same rules apply to them as to those already mentioned, except that the caesarean section is oftener indicated. Difficulty also not seldom occurs in women with normal pelves from an excessive size in the child through prolonged pregnancy, bigness of one or both parents, or the advanced age or multiparity of the mother. The child's head alone may be of excessive size. Some monsters offer difficulty in delivery from size or shape, but, of course, they are human beings, and are to be considered as such in delivery. The technique of the caesarean section has only a medical signification, and it need not be described here.

Symphyseotomy is an operation in which the joint of the pelvis at the symphysis pubis is cut, and the pelvis is allowed to gape so as to let out the child. The operation has fallen into disrepute. The mortality as regards the mother is about the same as in the caesarean section, but the mortality of the children is higher. In symphyseotomy the infantile mortality is about 9 per centum, while in the caesarean section it is practically nothing. If in symphyseotomy an error is made in estimating the size of the pelvis or the child's head—and such an error is often possible—the child will be killed, but in the caesarean section these errors make no difference. After the caesarean section the woman recovers promptly; after the symphyseotomy she recovers very slowly, and she may receive permanent injury.

Craniotomy is an operation wherein the head of the child is reduced in size to render delivery possible. The skull is perforated and the brain is broken up and removed or crushed out. Embryotomy is a similar operation wherein the viscera of the child are removed through an incision made in its thorax or belly (evisceration), or the head of the child is cut off (decapitation). There are numerous instruments and methods for performing craniotomy and embryotomy, but they all open the skull or belly, remove the brain or viscera, and then extract the child's body.

If the infant is hydrocephalic and is alive, the advocates of the operation warn us to be careful after opening the head to push the perforator into the base of the skull and stir it around well, so as to be sure the child will not be born alive. Pernice has recently reported a case of hydrocephalus which was delivered by craniotomy, but the operator did not work his perforator efficiently, and the child recovered, and grew up an idiot. A similar case occurred in Baltimore.

The indications for craniotomy among those that advocate its occasional use (and they are many) is in those cases in which the woman is so infected that caesarean section is dangerous, or where a child is hydrocephalic, or where an after-coming head is jammed (in this case even a caesarean section will not effect delivery), or in the case of a narrow pelvis and a moribund child, or finally in the practice of a country physician, who can not in an emergency get an assistant to do a caesarean section. One man can do craniotomy, but it requires three to perform the caesarean section. If the woman's narrow pelvis has a conjugata vera of five or more centimetres, craniotomy, if properly done, is not dangerous to the mother. With a conjugata vera less than 5 centimetres it is more fatal than the caesarean section. If the women are septic, the mortality in craniotomy is from 10 to 15 per centum; in caesarean section about 25 per centum.

As to the morality of craniotomy on the living or moribund child, it is not permissible under any possible circumstances: a consideration of the ethical principles set forth in the article on Ectopic Gestation will make this assertion clear.

The Congregation of the Holy Office on August 19, 1888, decreed that "In scholis catholicis tuto doceri non posse licitam esse operationem chirurgicam quam Craniotomiam appellunt." They gave a similar decision May 28, 1884, and they repeated the prohibition, with the papal approbation, on July 24, 1895. The text of these decrees may be found in the article on abortion, miscarriage, and premature labour.

The Porro operation consists essentially in a removal of the uterus after caesarean section to prevent further conceptions. As a means to prevent conception it is altogether unjustifiable, because repeated caesarean sections in the same woman, if the surgeon is at all competent, are practically no more dangerous than normal labour.

AUSTIN ÓMALLEY.

MATERNAL IMPRESSIONS

There is a wide-spread persuasion that a child, while carried in the womb of its mother, may be marked as the result of incidents that produce violent impressions upon her nervous system. This is so old a conviction in the human race and would seem to be substantiated by so much evidence that it is extremely difficult to convince people that there is no scientific basis for it. As a matter of fact, however, there is something mysterious about the way in which certain things that happen to the mother seem to affect the child in utero. As the result of the common belief in the truth of maternal impressions, mothers sometimes are prone to blame themselves for not having been sufficiently circumspect during the time of their pregnancy, and accordingly they may seek advice and consolation in the matter from clergymen. Women sometimes become very much depressed as a consequence of an unfortunate event of this kind, and as the simple truth is the best possible source of consolation, it would seem that a special chapter should be given to the subject in a work of this kind.

The evidence for the truth of the theory of maternal impression is almost entirely due to peculiar coincidences. James I. of England, the son of Mary Queen of Scots, could never stand, according to Sir Walter Scott, the sight of a drawn sword with equanimity, and it is said even that he nearly fainted at his coronation because of an unexpected glimpse of some naked blades in the hands of courtiers. This peculiarity was attributed to the fact that his mother, while carrying him in utero, had witnessed the violent death of her secretary, the unfortunate David Rizzio. There have been, however, any number of men who paled at the sight of a drawn sword before and since James I., with regard to whom no such circumstantial story could be told to account for it. There have been any number of women that have witnessed bloody murders under circumstances quite as heartrending as those surrounding Mary Queen of Scots and her secretary, and yet their offspring, though at the time in utero, have not been disturbed at the sight of drawn swords, nor of blood or any other circumstance connected with the deep impression that must have been produced on their mothers.

There is, of course, a striking instance related in the Old Testament, which seems to make it very clear that a belief in maternal impressions existed from the very earliest times among the Israelites. The story of Jacob is well known: "Jacob took him rods of green poplar and of the hazel and chestnut tree and pilled white streaks in them and made the white appear which was in the rods, and he set the rods which he had pilled before the flocks in the watering troughs when the flocks came to drink, and the flocks conceived before the rods and brought forth cattle, ring-streaked, speckled and spotted." In this case it seems evident that Jacob was not looking for a miracle, but was expecting that a law of nature would be fulfilled in the matter, the influence of the unusual sight upon the animal mothers proving sufficient to have a definite effect upon their unborn offspring. The most ardent advocates of the power of maternal impressions would scarcely concede the existence of as much influence as this of the mother's mind over the child unborn, otherwise there would surely be a very absurd collection of anomalous births in the race.

On the other hand, it is generally conceded that the mother's habitual temper of mind and the thoughts with which she occupies herself may influence her unborn offspring to a most marked degree. The story is told of a child-murderer who delighted in fiendish deeds of cruelty and had murdered many people in cold blood, that his mother, the wife of a butcher, had delighted in watching the operation of slaughtering during the course of her pregnancy. There are any number of women, however, who have, by the necessities of their occupation, had to witness the shedding of animal blood under such circumstances and yet without any special effect being noticeable in their offspring. It has been said that the opposite is also true, and that if a woman occupies herself with high and lofty thoughts, with noble deeds and unselfish devotion to others and if she occupies her mind and senses with the great works of art, a correspondingly beneficial effect will be noted upon the character of the foetus. These are, however, abstruse speculations leading to conclusions not founded upon actual observation, but upon theorising over the supposed fitness of things.

Coincidence plays such a large part in the matter of supposed maternal impressions that it is impossible to decide how much there is of fact and of consequence in the many stories that are told. Most women are a little afraid, as the time of their labour approaches, lest something or other— usually of an indefinite nature—that has happened during their pregnancy, may cause the marking of their child. When they find that the child is perfectly normal, they breathe a sigh of relief and forget all about it. If any anomaly is noted, however, then they are sure to connect it with some incident during pregnancy, and imagination is apt to lend details that confirm the supposed connection. On the other hand, there are not a few cases in which such anomalies have occurred, and good, sensible mothers have been unable to recall anything that might possibly serve to account for the peculiarity noticed in the child, though corresponding peculiarities in other children were supposed to be readily traceable to maternal impression. Even where there has been no foreboding of evil results, something or other that has occurred during the pregnancy will often be magnified enough by memory to account for the supposed maternal impression.

Doctors are very familiar with this tendency to make up stories to account for various deformities. It used to be considered that hip-joint disease and Pott's disease were the result of injuries in early life. They are now known to be due to tuberculous processes not necessarily and indeed only very seldom connected with injuries of any kind. Mothers are nearly always able to account in some way, however, for the beginnings of the disease in some accident that has happened. Young children are apt to have so many falls that some one of them is picked out as the probable cause of the disease that subsequently manifests itself in the joints. It is just this state of affairs that occurs with regard to supposed maternal impression. Some incident that would be otherwise unthought of is magnified into an accident that caused a serious nervous shock, and consequently led to the marking of the child.

In general it may be said for the clergyman's direction, that if women have, as is sometimes the case, a morbid sense of their guiltiness with regard to some maternal impression that has set a mark upon their child, such a state of feeling may very well be rendered less poignant by a frank statement of the present attitude of mind of most physicians with regard to the possible effects of maternal impressions. Scepticism is much more the rule than it used to be, and as time goes on fewer and fewer of the cases that used to be considered so inexplicable in the direct relationship that seemed to exist between maternal impression and deformity in the child are reported. Fifty years ago nearly all the authorities on this subject were agreed in considering that maternal impressions did play some part, though they could not explain just how, in the production of certain deformities. Now we venture to say that most of the thinking physicians who have occupied themselves with this subject would scarcely hesitate to say that they were utterly incredulous of any such effects being produced. The lack of any direct nervous or blood connection between mother and child is the basis for such disbelief, and is of itself the best argument against the old tradition.

With regard to mental defects, as a rule, not so much is said as for bodily defects. Bodily deformities are noted at once after birth, and then the mother recalls some incident of the pregnancy to account for them. Mental defects are, however, noticed much later, and are not so likely to be considered as connected with incidents of the puerperal period. There is no doubt that if the mother has had to pass through a series of emotional strains, or has suffered from severe shocks, children are likely to be born with diminished mental capacity. This is, however, not difficult to understand, since such incidents produce disturbances of the nervous system of the mother, and consequently also of her nutrition, and this is prone to be reflected in the child's condition, especially in that most delicate part of the child's organism, the brain. Hence it is that children born during the siege of Paris, or shortly after, were defective to such a marked degree that they were spoken of as "children of the siege," and this was considered to be quite sufficient explanation of nervous peculiarities later in life.

Baron Larrey, the distinguished French surgeon, made a report with regard to the children born after the siege of Landau in 1793. Of 92 children, 16 died at birth, 33 died within ten months, 8 showed marked signs of mental defects, most of them to the extent of idiocy, and two were born with several broken bones. In this case, however, it is well known that besides the shock of the danger consequent to the siege and the fear and distress of the women with regard to their husbands and relatives, there were added many privations and physical sufferings. The nutrition of the mothers was seriously disturbed by these, and it might well be expected that the children should suffer severely. The statistics of such events are not available in general, and when an effort is made to establish a cause for idiocy under other circumstances, none is usually found. Out of nearly five hundred cases of idiots whose histories were carefully traced in Scotland, in only six was there any question of maternal impressions having been the cause of the condition.

Of course there are many very wonderful coincidences that seem to confirm the idea that impressions made upon the mother's mind are sometimes communicated to the child in her womb. That they are not more than coincidences, however, is rather easy to demonstrate in most cases, since, as a matter of fact, at the time when the incident occurred which is supposed to have caused the deformity in the foetus, the stage of development of the intrauterine child has passed long beyond the period when formative defects could occur. For instance, it sometimes happens that the child-bearing woman sees an accident especially to the father of the child involving the loss of a limb. If, by chance the child should be born with a missing member, as sometimes happens, then there would seem almost to be no doubt of a direct connection between the accident witnessed, the effect produced upon the mother's mind, and the consequent deformity.

We know now that the formation of the limbs of the foetus is complete by the end of the third month. At this time the woman is scarcely more than conscious of the fact that she is pregnant, and it is not during this early period, as a rule, but during a much later period, that maternal impressions are supposed to have their influence. It is only such maternal impressions as occur very early in pregnancy, before the tenth week as a rule, that could possibly have any effect in the production of such deformities. It is by no means infrequent, however, to have children born lacking one or both limbs. Sometimes nothing but the stumps of limbs remain. In such cases it is now well known that intrauterine amputation has taken place. Some of the membranes that surround the child, especially the amnion, become separated into bands which surround tightly the growing members of the foetus and by shutting off the blood supply through constant pressure, lead to the dropping off of all that portion of the member lying below the band.

Not infrequently it happens that when a child is born thus deformed, the mother, by carefully searching her memory, can find some dreadful story that she has read, some accident that she has seen or heard of, and that has produced a seriously depressing effect upon her at the time, to which she now attributes the deformity that has occurred. Until the unfortunate appearance of her child was reported to her, she had no idea of any possible connection between the story and the bodily state of her intrauterine child. In not a few cases, however, the most faithful searching of the memory fails to show anything which could, by any possible connection, be made accountable for the deformity; and these cases, we may say at once, are in a majority.

Not a little of a popular notion with regard to the influence of maternal impression is due to the repetition of certain village gossip which by no means loses its point or effectiveness passing from mouth to mouth. On the other hand, maternal impressions have been exploited by novelists, who have found that the morbid curiosity of women particularly with regard to this subject may make their stories more widely read. Lucas Malet, who, in spite of the apparently masculine pseudonym, is really the late Rev. Charles Kingsley's daughter, has recently called renewed attention to this subject by her novel "Sir Richard Calmady." In this the hero is born with both his lower limbs missing from just below the knees. The author has been careful, however, with regard to the details of the supposed maternal impression to which this deformity is attributed. A young married woman in the early part of her first pregnancy has her husband, whom she loves very dearly, brought back to her with both his limbs taken off by a shocking accident which resulted fatally. It is not impossible, some physicians might think, to consider that so severe a shock could produce a very deleterious effect upon the foetus. That the result should so exactly copy the scene which was brought under the eyes of the young mother is, however, beyond credence. Occasionally such stories, supposedly on medical authority, find their way into the newspapers, usually from distant parts of the country. Certain parts of Texas particularly seem to be a fruitful source of such stories for newspaper correspondents when there is a dearth of other news. Farmers in thinly settled parts of the country lose a foot in a reaping machine or a hand in the hay-cutting machine when there is no one near to help them but their wives, with the result that the shock to their wives proves the occasion of a similar deformity in an as yet unborn child. Careful investigation of such cases, however, has invariably shown that either they were completely false or that the details showed that whatever had happened was at most a coincidence and never a direct causative factor in the subsequent deformity.

The greatest difficulty in the mind of the medical man, with regard to the possibility of maternal impression being communicated in any way to the foetus, is, as we have said, his knowledge of the anatomy of mother and foetus. While it is generally supposed that the mother is very intimately connected with her child in utero, the actual connection is by no means so direct as might be expected from the popular impression. It is usually considered that the mother's blood flows in the child's veins; but this is absolutely false. The child's blood is formed independently of the mother's blood quite as is that of the chick in the egg. At all times the blood of the child remains quite different in constitution to that of its mother. It contains many more red cells than does her blood and differs in other very easily recognisable ways. Mother and child are connected by means of an organ known as the placenta, which is attached very closely to the uterine wall and from which through the cord the blood of the foetus circulates. This placenta constitutes the so-called afterbirth. The mother's blood flows in one portion of it, that of the child in another, and they always remain distinct and separate from each other. The gases necessary for the child's life diffuse through the membrane which separates the two different bloods, and the salts and soluble proteids necessary for the child's nutrition, as well as the water necessary for its vital processes, all pass through this membrane, but at no time is there any direct blood connection between mother and child. Indeed, for a large part of the formative period of the foetus life, that is, during the first two months of its existence, the ovum is not very closely attached to the uterus at all, but grows by means of the vital power which it has within itself.

Nor is there any direct nervous connection between mother and child; indeed, there are no nerves at all in the placenta, and none in the cord through which all communications between mother and child must pass. It seems impossible to explain, then, how maternal impressions can so effectively pass from mother to child; and indeed, the whole subject, when looked at in this way, is apt to be considered legendary, and the facts adduced in support of the theory of maternal impressions are practically sure to be thought mere coincidences. A little knowledge here might seem to justify many things that more complete knowledge fails to be able to find any reasons for.

There is no doubt, however, that the mother's environment during pregnancy is in general very important for the perfect development of the intrauterine child. Many more deformed births are reported after times of stress and trial, as, for example, after the sieges of great cities, notably the siege of Paris in 1871, and such scenes of desolation as occurred during the thirty years' war in Germany. These are, however, not direct, but indirect effects of maternal impressions. The development of the human being in utero is an extremely complicated process. Any disturbance of it, however slight, is sure to be followed by serious consequences. Disturbances of nutrition, such as are consequent upon the deprivation that has to be endured in times of war or during sieges, is of itself sufficient seriously to disturb even the uterine life of the child. In these cases, however, there will be no traceable connection between the form of the maternal impression and the type of deformity that occurs. This is, however, the essence of the old theory of the direct effect of maternal impressions, and consequently that theory must fall to the ground.

From all that has been said, however, it becomes very clear that as far as possible women should be shielded from the effect of various nervous shocks during their pregnancy, and that they owe it to themselves and their offspring to be careful with regard to any morbid manifestations of feeling that they may detect in themselves.

JAMES J. WALSH.

HUMAN TERATA AND THE SACRAMENTS

Teratology (, a monster) is a part of biology that treats of deviation from a normal development in man and the lower animals. The name was adopted in 1822 by the elder Saint-Hilaire, who then attempted to separate the results of modern exact methods of research from the myths and loose descriptions of monsters found in the writings of old authors. Cicero (De Divinatione) derives the term monster from the proper preternatural signification looked for in the occurrence of these abnormal beings: "Monstra, ostenta, portenta, prodigia appellantur, quoniam monstrant, ostendunt, portendunt et predicunt."

At the end of the seventeenth century Malpighi and Grew discovered that plant tissue is entirely made up of microscopic spaces enclosing fluid; they called these spaces cells. Different investigators found that animal tissue is also composed of cells; and between 1835 and 1839 Schwann and Schleiden formulated the law that every metazoic organism is made of cells, and starts from a cell.

In 1672 de Graaf discovered the mammalian ovum, in 1675 Ludwig Ham found spermatozoa, in 1827 von Baer recognised the human ovum, but not until 1875 was the important fact established that fertilisation is effected by the fusion of the male and female pronuclei. This was demonstrated by Oscar Hertwig from observation of the ova of starfishes.

Mammalian ova, owing to an almost complete lack of yolk, are all small. The egg of a whale is about the size of a fern-seed, but the yolked eggs of birds are large—that of the great auk was 7.5 inches long. In man the ovum is from 0.18 to 0.2 mm. in diameter, scarcely visible to the naked eye, and the spermatozoon is extremely minute. The human spermatozoon is only fifty-four thousandths of a millimetre in length, and from forty-one to fifty-three thousandths of a millimetre are taken up by its flagellum. The essential part is from four to six thousandths of a millimetre in length (Dr.

L. N. Boston, Journ, of Applied Microscopy, vol. iv. p. 1360). A line of 18 human spermatozoa would reach only across the head of an ordinary pin. These spermatozoa have the power of locomotion in alkaline fluid. Henle found they can travel one centimetre in three minutes.

The human ovum and spermatozoon are single cells, and the principal parts of a typical cell are the cytoplasm (called also the protoplasm), and, within this, the nucleus and centrosome. The centrosome is efficient in the process of cell-division. A few cells have also an outer envelope or membrane, and this part is well developed in the ovum.

The nucleus is the centre of activity in a cell. In the resting state it is surrounded by a membrane, and within the membrane is an intra-nuclear network made up of chromatin and linin—the chromatin is an important element. The meshes of this network are probably filled with fluid.

During the stages preparatory to the mitotic, or indirect, division of a cell into two cells (one of the methods of reproduction) the chromatin segregates in typical cases into two groups of loops, and each group has equal portions of the chromatin. When the chromatin is in this shape, a loop is called a chromosome.

The chromosomes are very important. They occur in constant definite numbers in the somatic cells of the various species of many animals and plants, and it is probable that each species of plant and animal has its own characteristic number of chromosomes. Wilson (The Cell in Development and Inheritance, New York, 1890) gives a list of 72 species in which the number has been determined. Man has probably 16 chromosomes in the somatic cell, and the mature male and female germ cells in man contribute eight chromosomes each to the nucleus of the impregnated ovum.

The chromosomes transmit the physical bases of heredity from one generation to the next, and the heritages from the two parents are equal except in cases of prepotency. Every cell in the human body is derived from the father and the mother equally. The fact that the woman carries a child for months in her womb means only that she employs a peculiar method of feeding and protecting it. After its birth she feeds it from her breasts, before birth through its umbilical vessels, but she originally gives only the eight chromosomes as the father does, and the child's vital principle builds up the body from this foundation. The popular notion that the foetus in the womb is formed through some process of literal abstraction from the maternal tissues is no more true than that the infant is so built up while it is suckling; both processes are merely different methods of feeding.

All the chromosomes from the fathers of at least 200 men could fit simultaneously on the head of one pin, yet virtually, not merely potentially, half the bodily substance of that multitude, and all the physical characteristics derived from the 200 fathers, are indubitably contained in those chromosomes and nowhere else, unless by a special creation they are infused with the new soul, which seems to be an altogether unreasonable alternative. This statement concerning the minuteness of the chromosomes is not speculation—they can readily be seen and measured with the aid of the microscope.

A human being, then, obtains eight microscopic chromosomes from his father and eight from his mother, positively nothing more except food; yet he develops into a man with a body made up of countless millions of cells which expand into more than 200 bones in the skeleton and over 200 muscles,—into the fascias, ligaments, tendons, the great and small glands, the lymph and blood systems, the respiratory and alimentary tracts, the skin and its appendages, and a nervous system, which alone furnishes material for years of study if we would learn its anatomy fully. Not only all this, but the man commonly closely resembles his father or his mother, or some other ancestor, in personal appearance, in certain physical tendencies, in graces or blemishes; and furthermore, he shows inherited racial characteristics.

If a father is prepotent, he may have a greater effect in producing the formed child than the mother has, and vice versa, as when a son closely resembles his father or his mother. Prepotency, moreover, may extend down through generations and centuries. In the streets of Palermo to-day typical Normans may be seen, despite the intermarriages of centuries, who are the descendants of those male Normans that went down to Sicily with Tancred. There are Romans there, too, and Saracens. When the Belgae—a race of tall, red-bearded men, with elliptical skulls—went from the continent of Europe to Ireland, probably six centuries before our era, they conquered the aborigines, a gentle, brune race of lower stature. These Belgae became the ancestors of the chieftain class, and their physical type persists until to-day; so does that of the Pictish aborigines. Daniel O'Connell had a typical Belgic body. Other big, blond Irishmen are Norse or Danish in remote origin.

How is the extremely complex human body with its various physical characteristics built up from the nucleus of a fecundated cell, the ovum? The endeavour to answer this question has brought out most ingenious speculation from nearly all the great biologists of modern times. The question is the foundation of the theories of heredity, and it is also fundamental in the theories of evolution.

The human ovum is a flattened spherical cell, made up of a very delicate cell-wall, called the vitelline membrane; outside this is a comparatively thick membrane, the zona pellucida, which is properly not a part of the cell. Within the vitelline membrane is a granular cytoplasm, the vitellus (yolk), and in this lies the nucleus, which in the old text-books was called the germinal vesicle. This nucleus contains a nucleolus.

The human spermatozoon consists of a flattened head which has a thin protoplasmic cap extending down two-thirds of its length. In the head is the nucleus with the chromatin. Beyond the head is the neck, which contains the anterior and posterior centrosomes. Behind the neck is the tail, or flagellum, in three parts,—the middle piece, the principal part, and the end piece. From the neck to the end of the tail centrally runs a bundle of fibrils, the axial filament. In the middle piece these fibrils are wrapped within a single spiral filament which winds from the neck down to the annulus at the beginning of the principal part, and lies in a clear fluid. Without the spiral filament, along the middle piece, is the mitochondria, a finely granular protoplasmic layer. The principal part of the tail consists of the axial filament enclosed in an involucrum, and the end piece is made up of this filament without the involucrum.

The head and neck of the spermatozoon, which contain the nucleus and centrosomes, are the essential parts, and the middle piece and the remainder of the tail appear to be used solely for locomotion and penetration. When the head penetrates the ovum, the tail is detached and rejected.

Our knowledge of the initial stages in the development of a human embryo is derived indirectly from the observation of other mammals. There are nine early human embryos reported, and the average probable age of these is twelve days. Breuss' specimen was probably ten days old (Wiener med. Wochenblatt, 1877). Peters (Einbettung des mensch. Eies, 1899) found a smaller embryo than this. The Breuss ovum was 5 mm. in length; Peters' was 3 by 1.5 by 1.5 mm., but the probable age was not given. There have been numerous embryos more than twelve days old observed, and since the process after the twelfth day is identical in man and the higher mammals, there is no doubt that the first stages are also the same.

The segmentation that makes new cells is complicated, and the outcome of the division is a ball of cells. In eggs which have a large yolk, like those of birds, the cells form a round body resting on the surface of the yolk, but in mammalian ova a hollow ball of cells, or a Morula, results, which lines the internal surface of the cellular envelope. The ovum absorbs moisture by osmosis and enlarges, and about the twelfth day after the germ-nuclei have begun to divide, the Morula, or hollow ball of cells, called also the Blastodermic Vesicle, is formed.

The next stage in development is the establishment of two primary germinal layers, called together the Gastrula, The outer layer is the Ectoderm or the Epiblast, and the inner layer is the Endoderm or Hypoblast. In a Morula the smaller cells, which contain less yolk-material, gradually grow around the larger yolk-containing cells to form the Gastrula.

Between the Ectoderm and the Endoderm a layer of cells called the Mesoderm or Mesoblast is next formed, and from these three layers all the parts of the embryo are built up. From the outer Ectoderm and the inner

Endoderm those organs arise which are in the body, outer and inner,—as the nervous system and the outer skin from the Ectoderm, the inner entrails, the lungs and liver, from the Endoderm. From the Mesoderm come the inner skin, the bones and muscles.

By this time the embryo is a minute longitudinal streak at the surface of one pole of the ovum. The "Primitive Trace" is like a long inverted letter U, the legs of which are in apposition. The Primitive Trace becomes a circular flattened disc; and it grows into a cylindrical body by the juncture of the free margins which fold downward and inward and meet in the median line, and this closes in the pelvic, abdominal, thoracic, pharyngeal, and oral cavities. The legs and arms bud from this cylinder later. While the ventral cylinder is growing, another longitudinal cylinder is formed along the upper surface of the embryo, which will contain the brain and the spinal column. The subsequent development of the embryo and foetus need not be known for an understanding of the material considered in treating here of terata.

Human terata occur in certain rather definite, types of erroneous development, and the classification of Hirst and Piersol (Human Monstrosities, Philadelphia, 1891), which is a combination and change of the classifications of Geoffrey Saint-Hilaire, Klebs, and Förster, is the most satisfactory. There are four great groups of abnormally developed human beings: (1) Hemiteratic; (2) Heterotaxic; (3) Hermaphroditic; (4) Monstrous.

Hemiterata are giants, dwarfs, persons showing anomalies in shape, in colour, in closure of embryonic clefts, in absence or excess of digits, or having other defects. This group does not come under discussion here, but attention should be called to the fact that women who are dwarfs are to be warned before marriage that they cannot be delivered normally,—that the caesarean section or symphyseotomy will be necessary, or that certain physicians will practise craniotomy in delivering them.

The Heterotaxic group comprises persons whose left or right visceral organs are reversed in position through abnormal embryonic development; the liver is on the left side, the heart points to the right, and so on.

Of the next group, the Hermaphroditic, it may be said that a true hermaphrodite, in the full sense of the term, has not been found; but there have been several examples of individuals who had an ovary and a testicle, and other rudimentary sexual organs that belonged to both male and female. Forms of apparent doubling are common, and in case of doubt as to sex the probability leans toward the masculine side. As to marriage in such cases, questions may arise that are to be settled by the anatomist. In dealing with double monsters it is sometimes difficult or impossible to determine whether we have to do with one or two individuals, and this difficulty has serious weight, especially in the administration of baptism. It is improbable that there is a doubling of personality in hermaphrodites. A striking characteristic of compound terata is that the individuals are always of the same sex; moreover, the embryonal development of reproductive organs in general is such as almost to preclude a question of duality of personality.

Terata, more properly so called, are divided into single, double, and triple monsters. Single monsters may be autositic, or independent of another embryo or foetus; or they may be omphalositic, that is, dependent upon another embryo or foetus, which is commonly well developed, and which supplies blood for both through the umbilical vessels. When an omphalosite exists, the other foetus is called, in this case also, the autosite.

The first order of autositic single monsters contains four genera with eight species, and under these species are thirty-four varieties. They may have imperfect limbs, no limbs, one eye in the middle of the forehead (cyclops), fused lower limbs (siren), and so on. Some of these monsters show a strong resemblance to lower animals, but there is no record that is in any degree scientific of a hybrid between a human being and a lower animal.

There are two genera of the omphalositic single monsters, with four species. One of the twins, the autosite, is commonly a normal child; the other, the omphalosite, may be as small as a child's fist, and be very much deformed. Of these omphalosites the paracephalus has an imperfect head, commonly no heart, and the lungs are absent or rudimentary. The acephalus has no head, and commonly no arms; the asomata is a head more or less developed, with a sac below containing rudiments of the trunk organs. The Acephalus is very rare—the rarest of all monsters except the Tricephalus. There is a fourth kind—the foetus anideus. This is a shapeless mass of flesh covered with skin. There may be a slight prominence with a tuft of hair on it at one end of the mass to indicate the head. In this monster there are more traces of bodily organs than might be expected. These four kinds of omphalosites are either dead when born, or they die as soon as the placental circulation is cut off. If there is any probability of life, the physician should give them baptism before the placental circulation is stopped.

Nothing satisfactory is known concerning the etiology of single monsters. Landau, and other authorities as great as he is, reject the theory that maternal impressions from fright or exposure to the sight of hideous deformity are the cause of terata. I think the father is accountable for terata as often as the mother is. Barnes, an English physician, and others claim they find that terata are frequent in consanguineous marriages, but I have not been able to verify the assertion.

It seems a theory may be offered to explain the single terata. In 1888 Roux of Breslau by puncturing one blastomere of a frog's egg in the two-cell stage killed the punctured blastomere without affecting the other. The punctured blastomere remained inactive, but the other developed into a complete half embryo.

Crampton by separating and isolating the blastomeres in the two-cell stage obtained a half embryo; and Zoja by isolating blastomeres of the medusae, Clytia and Laodice, got dwarfed larvae.

Wilson succeeded by the separation through shaking of the blastomeres in the two-cell and four-cell stages in developing Amphioxus larvae, which were half the natural size for the two-cell blastomeres, and commonly half the normal size from the four-cell blastomeres, yet in the latter some of the larvae were of the normal size but imperfect. From the eight-cell stage he got only imperfect larvae. Similar results were obtained by other operators with various eggs.

Driesch and Morgan by removing part of the cytoplasm from a fertilized egg of the ctenophore, Beroe, produced imperfect larvae showing certain defects which represent the parts removed.

In these cases of injured and isolated blastomeres we have, it seems to me, a plausible theory for the etiology of single terata. The blastomeres in the human ovum may perhaps be injured in part by toxins from the mother, or they may be defective through disease in the ovum or the spermatozoon. They also may possibly be displaced traumatically, but this seems to be doubtful.

There are three theories concerning the origin of omphalositic terata. Ahlfeld (Missbildungen des Menschen, Leipsic, 1882) holds that the autosite is stronger than the omphalosite, and as a consequence the foetal circulation in the omphalosite is reversed, and development is thus checked. Dareste (Production artificielle des monstruosités, Paris, 1876), Panum (Beitrag zur Kenntniss der physiol. Bedeut. der angeboren Missbildungen, Virchow's Archiv., 1878), Perls (Lehrbuch der allgem. Pathologie) and Breus (Wiener med. Jahrbuch, 1882) maintain there is an inherent original defect in the omphalositic child which prevents development of the blood-vessels, and that Ahlfeld's theory of an indirect umbilical connection of the omphalosite to the placenta is not probable; if it were, omphalosites would be very common, because one of twins is nearly always stronger than the other. Hirst and Piersol (op. cit) combine these theories. This kind of monster is certainly an imperfectly developed human individual, and even the Foetus Anideus should receive at the least conditional baptism.

The next group comprises the composite monsters. Normal twins may arise from the fertilisation of one ovum and of two distinct ova. In 506 cases examined by Ahlfeld he found that 66 twin births came from single ova. Twins from a single ovum are always of the same sex, and they are not easily distinguished one from the other. Triplets may arise from one, two, or three ova. The elder Saint-Hilaire thought that composite monsters arise from the fusion of two impregnated ova, but this opinion is now generally rejected. Composite terata in every instance arise from a single ovum.

There is a divergence of opinion, however, as to the origin of a composite monster in the single ovum. Some authorities maintain that these monsters arise from the union of two originally separate primitive traces. This supposes primitive duality followed by fusion (Verwachsungstheorie). Other writers hold that there is originally one primitive trace, and that composite terata are the product of a more or less extensive cleavage of this single blastoderm. This supposes primitive unity followed by fission (Spaltungstheorie). Here, as in the case of normal development, the argument is founded on analogy. The earliest stage in the development of a human double monster observed was at the fourth week after fertilisation—Ahlfeld's case.

B. Schultze (U. anomale Duplicität der Axenorgane, Virchow's Archiv.) and Panum and Dareste (op. cit.) hold the fusion theory— the fusion of two separate blastoderms in one ovum. Panum and Dareste have seen two separate normal blastoderms on one ovum. Allen Thompson in 1844 (London and Edinburgh Monthly Journal of Medical Science), Wolff, von Baer, and Reichert also observed two embryos in one ovum. Dareste is of the opinion that the fusion of two separate ova is impossible. The fission theory—the fission of a single blastoderm to make a composite monster— is supported by Wolff, J. F. Meckel, von Baer, J. Müller, Valentine, Bischoff, and others, especially by Ahlfeld. Ahlfeld says that this single blastoderm is split by pressure.

Gerlach also (Die Entstehungsweise der Doppelmissbildungen, etc., Stuttgart, 1882) admits fission, but he contends that it is not so simple a process as Ahlfeld thinks it is. It is not a passive cleavage, but a result of a force in the cell-mass existing before differentiation. Gerlach calls fission at the anterior or head-end of the single blastoderm, bifurcation; and he has actually observed such bifurcation in a chick embryo of sixteen hours (U. d. Entstehungsweise der vorderen Verdoppelung. Deutsche Archiv. f. klin, Med., 1887). In this case the first change noticed was a broadening of the anterior end of the primitive streak; next a forked divergence appeared, and this became more pronounced; until by the twenty-sixth hour the bifurcation was half as long as the undivided posterior part. From each anterior end of the diverging branches a distinct head-process extended. Allen Thompson (loc. cit.) in 1844 saw a goose-egg, which had been incubated for five days, in which was a double monster divided to the neck.

Beyond this observation by Gerlach we have the fact, which seems to make for the fission theory, that no matter how unequally nourished or how variable in extent, the union between the halves of double monsters is always symmetric—exactly the same parts of each twin are joined. This seems to exclude a fortuitous growing together of dissimilar areas or cell-masses, for non-parasitic double terata at the least. Born (U. d. Furchung des Eies bei Doppelbildungen, Breslauer Aerztl. Zeitschr., 1887), in a study of fish ova, found that ova which produce double monsters begin with a segmentation like that of the single normal ovum.

If fission is complete homogeneous twins are the result; these twins are of the same sex and very similar in appearance. Incomplete fission, as has been said, gives rise to double or triple terata. If one of the teratic twin embryos is stronger than the other, the various combinations of enclosure and parasitism may result, although the origin of parasitic double terata is not convincingly clear. A triple monster, according to the fission theory, arises from a double incomplete cleavage of the primitive trace. Dr. Ephraim Cutter has observed teratic composite spermatozoa which, he thinks, probably have influence in producing composite monsters.

There are three orders of the double autositic monsters: Terata Katadidyma, in which the embryonal fission was at the cerebral end; the Terata Anadidyma, divided below; the Terata Anakatadidyma, divided above and below, but joined at the middle of the body. There are four genera of the Terata Katadidyma with many species. The first genus is the Diprosopus, the double-faced. The doubling varies from the finding of two complete faces to a slight trace of duplex formation in one head. Förster in 500 human monsters observed 29 cases of diprosopi.

There are six species of diprosopi: 1. D. Diophthalmus, which has only two eyes, but there is a doubling of the nose. 2. D. Distomus, which has two mouths, two lower jaws, two tongues, one pharynx, and one oesophagus. 3. D. Triophthalmus, which has three eyes, and the doubling of the face is more complete. There are only two ears. 4. D. Tetrophthalmus, which has four eyes and two well-separated faces. 5. D. Triotus is like the last, but it has three ears. 6. D. Tetrotus has four ears, four eyes, and there is some doubling at the pharynx. Two oesophaguses enter one stomach in this species commonly. D. Tetrotus is rare—only one example in man is known. In all diprosopi there is only one trunk, one pair of arms, and one pair of legs. Sir James Paget had a photograph, made in 1856, of a living diprosopus, the second face of which had a mouth, nose, eye, part of an ear, and a brain (?) of its own. The two faces acted simultaneously, suckled, sneezed, yawned together.

Are diprosopi twins? An answer to this question will be clearer after a description of other composite terata.

The second genus of the Terata Katadidyma is the Dicephalus. This genus comprises five species, which have in each case two heads, with separate necks commonly. There are two vertebral columns, which usually are separate down to the sacrum, and they converge at the lower end. In the interior organs doubling will be found corresponding to the degree of separation of the trunks. In all the species of this genus there are one umbilicus and one cord.

The first species of the Dicephalus is the Dicephalus Dibrachius—a two-armed, double-headed monster. In this species most of the viscera are single, but the right and left halves of each viscus are supplied by the respective foetuses, and the entrail does not become indistinguishably single until near the lower end of the ileum. There may be two ordinary kidneys and a third smaller one, two pancreatic glands, and two gall-bladders. Such a monster may be monauchenous or diauchenous.

The next species is the Dicephalus Tribrachius Dipus—two heads, three arms, and two legs. There is also a Dicephalus Tribrachius Tripus (three arms and three legs), D. Tetrabrachius Dipus (four arms and two legs), and D. Tetrabrachius Tripus (four arms and three legs). In all these cases there is no doubt of the presence of twins, unless there might be some doubt as to dual personality in the Dicephalus Dibrachius. In the Dicephalus Tetrabrachius Dipus and the Dicephalus Tetrabrachius Tripus there is almost complete duplication of the internal organs, and the halves of the composite body belong evidently to individuals distinct in thought, volition, and character. Each brain controls only its own half of the body. There are four lungs, two hearts (sometimes in one pericardium), two stomachs, two intestinal canals down to the colon or lower, two livers (sometimes joined), four kidneys (or three, one of which is small), two bladders, emptied at different times through a common urethra.

Dicephali are somewhat common. Förster found 140 among 500 specimens of monsters. They are rarely born alive. The best known cases of dicephali that lived for any length of time are:

1. Peter and Paul, of Florence, born in 1316, lived thirty days.

2. The Scotch Brothers, born in 1490, lived twenty-eight years. They were at the court of James III. Above the point of union the twins were independent in sensation and action, but below the point all sensation and action were common. One died before the other, and the second "succumbed to infection from putrefaction" a few days later.

3. The Würtemberg Sisters, born in 1498.

4. The twins, Justina and Dorothea, born in 1627, lived six weeks.

5. Boy twins at Padua, born in 1691, lived to be baptised.

6. Rita-Cristina, born at Sassari in Sardinia in 1829. They lived eight months. These children had a common trunk below the breast, one pelvis, and one pair of legs. Rita was feeble and quiet, Cristina vigorous and lively. They suckled at different times; and sensation in the heads and arms was individual, but below the junction it was common. Rita died of bronchitis, and during Rita's final illness Cristina was healthy; but when Rita died, Cristina, who was suckling at the time, suddenly expired. They had two hearts in one pericardium, the digestive tracts did not fuse until the lowest third of the ileum was reached. The livers were fused, the vertebral columns were distinct throughout. These twins were baptised separately.

7. Marie-Rose Drouin, born in Montreal in 1878. They lived seven

months. Marie died of cholera infantum; and Rose then died, although she had not been directly affected by the disease. These twins were like Rita-Cristina anatomically except that they had no legs. The respirations and heart-pulsations differed, and one child slept while the other child cried.

8. The Tocci boys, born in Turin in 1877. In 1882 they were strong and healthy, and they may be living still. They resembled Rita-Cristina anatomically in every respect. Each boy had control of the leg on his own side, but not of the other leg, consequently they could not walk. Their sensations above the juncture were distinct, and their thoughts and emotions differed.

In the Paris L'union médicale there is an account of a bicephalic still-born monster, born at Alexandria in 1848, which, according to the report, had on one side a typical negro head and on the other side a typical Egyptian fellah head. This report is probably not authentic; but if it is, it would be difficult to reconcile it with the fission theory. Supposing the report true, the case would have to be one (1) of superimpregnation wherein (2) a spermatozoon from each source penetrated the same ovum, (3) a bicephalic monster resulted, with (4) distinct racial characteristics. All this is extremely improbable.

Superimpregnation has happened. There are cases where negresses have given birth to twins, one of which was a negro and the other a mulatto. Instances are cited in books on Legal Medicine like those of Tidy and Beck. In Flint's Physiology a case is recorded in which a mulatto woman in Kent County, Virginia, married to a negro, gave birth to twins, in 1867, one of which was a negro much blacker than the mother, and the other a white child, with long, light, silky hair, and a "brilliant complexion." The white child's nose was shaped like the mother's, but there was no other resemblance. Even supposing this to be a case of superimpregnation, that does not fully explain the extreme whiteness of one child and the extreme blackness of the other.

Superfoetation is also possible. Tidy (Legal Medicine) gives a case: "Mary Anne Bigaud, at thirty-seven, on April 30th, 1748, gave birth to a full-term mature boy, which survived its birth two and a half months, and to a second mature child (girl) on September 16th, 1748, which lived for one year." The second child was born four and a half months after the first, and both were "nine-months" children. It was proved after death in this case that the mother had not a double uterus, and the report is vouched for by Professor Eisenman, and by Leriche, surgeon-major of the Strasburg Military Hospital. Several other cases of superfoetation are given by Bonnar (Edin. Med. Journ., January, 1865).

The third genus of Terata Katadidyma is the Ischiopagus. These twins are divided so much from above downward that the heads are at almost opposite ends of the double body. They are joined at the coccyges and sacra, and the spinal columns have nearly the same axis. The trunk organs are complete and separate, except that they are commonly fused in the pelvis. There may be two, three, or four legs, given off at right angles to the pelvis. This kind of monster is not rare. Förster collected twenty cases, and nine new examples were reported in the Index Medicus between 1879 and 1893. Ischiopagic twins were born in County Roscommon, Ireland, in 1827, and baptised separately. The Jones Twins, born in Typhon County, Indiana, in 1889, lived for about two years; they were ischiopagi, and they had the very unusual quality, it is said, that they differed in complexion and the colour of eyes and hair. A case was reported in American Medicine, September, 1903.

Classed with the Katadidyma is the genus Pygopagus, although it has four legs. This form is very rare. The twins are joined only by the latero-posterior aspects of the sacra and coccyges, so that the two individuals are placed almost back to back. The trunk organs are independent, except for some fusion near the point of juncture. Examples of this class are the Hungarian Sisters, born at Szony in 1701, who lived to womanhood; the negresses Millie-Christine, born in 1851, and who were recently living in North Carolina; and the Blazek Sisters of Bohemia. The negresses had common sensation in the legs, but Millie could not localise what part of Christine's legs was touched, and vice versa.

The second group of the double autositic monsters are Terata Anadidyma—terata divided from below upward. The first genus is the Dipygus. This has a single body above, but a double pelvis with double lower extremities in the typical cases. There is an exact description of a double monster of this kind in the Gaelic Annals of the Four Masters as early as the year 727 of this era. The chronicler says in that year on Dalkey Island near Dublin, "There was a cow seen which had one head and one body as far as her shoulders, two bodies from her shoulders hindward, and two tails. She had eight legs, and she was milked three times a day."

A perfect human Dipygus with two equally developed pairs of legs is unknown. Catherine Kaufmann, who was born in 1876, and who died in 1878, had a double pelvis with double pelvic organs in part, but she had only one pair of legs. There is a similar anomaly said to be living in Philadelphia at present. Blanche Dumas, born in 1860, had a double pelvis, double pelvic organs, and three legs. Mrs. B., born in 1868, had four legs— the two inner ones were smaller than the outer pair. Her spinal column was divided up to the third lumbar vertebra. Her double pelvic organs acted independently. There are living male examples of this form of monster.

The next genus is the Syncephalus, called also Janus and Janiceps. Its lower body is double up to the umbilicus, the trunk single above that point; the head shows signs of doubling, and there are four legs and four arms; the bodies grow front to front. The head usually is large, therefore this monster is born dead.

Another genus is the Craniopagus—twins joined only by the skull or scalp. There are three species, named from the place of union—Craniopagus Frontalis, C. Parietalis, and C. Occipitalis.

A third group of double autositic monsters are the Terata Anakatadidyma, which are divided above and below, but joined from the navel to the head. There are three genera. The first, the Prosopothoracopagus, is joined at the upper abdomen, the chest, and the faces; the spinal columns are separate. The faces are imperfect, the jaws are united; there is a broad neck with one oesophagus, and there is one stomach and one duodenum. This is a rare form, and it can not exist out of the uterus.

A second genus, the Thoracopagus, has a thorax in common, and the inner legs may be united. It is, as a rule, still-born.

The next genus is the Omphalopagus, in which the twins are joined from the navel to the bottom of the chest. This double monster has the slightest union of all, and it is very rare. The Siamese Twins were omphalopagi. They quarrelled; one became a drunkard and the other remained temperate. They married two women, and Chang had ten children, and Eng twelve. Chang died while Eng was asleep, and the latter died two hours after he had waked and learned of his brother's death.

There is a genus, the Rachipagus, the examples of which are joined behind like the class Terata Anakatadidyma that are joined in front.

Four known attempts have been made to separate double monsters surgically, but all failed owing to crude surgery; modern methods might be successful in some cases.

The second order of double monsters comprises the parasitic class. There are three genera of these terata, with five species and seventeen varieties. The chief of these only will be mentioned. The Heterotypus is a parasitic child which hangs from the abdominal wall of the principal subject. Varieties of this species are the Heteropagus, which is a parasite with head and arms; the Heterodelphus, which has no head; the Heterodymus, which has a head, neck, and thorax. The Heteralitis is a second species, in which the parasite is inserted at a distance from the navel of the autosite. The Epicomus is the only example, and it consists of a parasitic supernumerary head. The Polypnathus is a parasite attached to the jaw of the autosite. When fastened to the upper jaw, it is an Epignathus; at the lower jaw it is an Hypognathns. Another group is made up of terata having parasitic legs which are attached to different parts of the autosite,—to the pelvis, the head, the abdomen, and so on. Finally, there is the Endocyma, which is a parasite enclosed within the body of an autosite.

Parasites are nourished through the blood supply of the autosite, and the parasites usually are incapable of motion. The autosite can feel when the parasite is touched, and in some cases the autosite can localise the touch. In India, in 1783, a child was born which had a supernumerary head attached to the autositic head, crown to crown; it lived four years. The parasite's eyes were always partly open, but they appeared to be incapable of intelligent vision. They contracted under strong light, and when the autosite was suddenly awakened both sets of eyes moved.

Gould and Pyle (Anomalies and Curiosities of Medicine) give an account of an Italian boy, aged eight years, who had a small parasitic head protruding from near the left third rib. Sensibility was common. Each of the heads received baptism (one was called John and the other Matthew), and there was question as to whether extreme unction should be administered to the parasitic head. A similar case occurred in England in 1880 (British Med. Journal), and the parasitic head could be pinched without attracting the attention of the autosite.

Teratologists now exclude Dermoid Cysts from the lists of terata. The hair, teeth, and particles of bone found in these cysts are looked upon as the development of abnormal ectodermic and endodermic cells, rather than as evidence of a separate personality.

There is only one well-authenticated case of a triple monster, and this happened in Italy in 1831. The monster had a single broad body with three distinct heads and two necks. It was killed in delivery.

In Katadidyma (terata divided from above downward), when we have dicephali, ischiopagi, or pygopagi, there are evidently two individuals present. Is the Diprosopus, however, the two-faced monster, possessed of one or two souls? The cases vary, as we said, from examples with two distinct faces and four ears to cases that have merely two noses. What portion of a human body is required to contain a new soul? That is an interesting question for the psychologist and a very practical one for the moralist, and no moralist has yet attempted to solve it. The presence of a brain is not essential, because acephalous monsters develop without brain, and they are born alive; they have a vital principle which is identical with the soul.

Among the Terata Anadidyma (divided from below upward) the Syncephalus and the Craniopagus are unquestionably two persons. Is the Dipygus (single down to the navel, double below) one or two persons? Mrs. B., the example already given, was as double below the navel as any Dicephalus is above that point. She had features so well ordered in unity that she was a pretty woman, but that unity ceased at her waist. Was her husband unknowingly a bigamist? I think he was. After a consideration of the fission of terata, and the non-essential quality of the brain, why should fission that started at the feet differ from fission that started at the head?

In the Rituale Romanum Pauli V. (tit. ii. cap. i. nn. 18, 19, 20, 21), the following directions for the baptising of terata are given:

18. In monstris vero baptizandis, si casus eveniat, magna cautio adhibenda est, de quo si opus fuerit, Ordinarius loci; vel alii periti consulantur, nisi mortis periculum immineat.

19. Monstrum, quod humanam speciem non praeseferat, baptizari non debet; de quo si dubium fuerit, baptizetur sub hac conditione: Si tu es homo, ego te baptizo, etc.

20. Illud vero, de quo dubium est, una ne, aut plures sint personae, non baptizetur, donec id discernatur: discerni autem potest, si habeat unum vel plura capita, unum vel plura pectora; tunc enim totidem erunt corda et animae, hominesque distincti, et eo casu singuli seorsum sunt baptizandi, unicuique dicendo: Ego te baptizo, etc Si vero periculum mortis immineat, tempusque non suppetat, ut singuli separatim baptizentur, potent minister singulorum capitibus aquam infundens omnes simul baptizari, dicendo: Ego vos baptizo, in nomine Patris, et Filii, et Spiritus sancti. Quam tamen formam in iis solum, et in aliis similibus mortis periculis, ad plures simul baptizandos, et ubi tempus non patitur, ut singuli separatim baptizentur, alias numquam, licet adhibere.

21. Quando vero non est certum in monstro esse duas personas, ut quia duo capita et duo pectora non habet distincta; tunc debet primum unus absolute baptizari, et postea alter sub conditione, hoc modo: Si non es baptizatus, ego te baptizo in nomine Patris, et Filii, et Spiritus sancti.

AUSTIN ÓMALLEY.

SOCIAL MEDICINE

The influence of the clergyman or the charitable visitor in matters of health and sanitation can scarcely be overestimated. The removal of prejudices with regard to sanitary regulations for the prevention of disease and modern advances in the treatment of disease is an important social duty. There is no doubt that if this influence be properly directed, sanitary measures of various kinds will be much more readily enforced and the precautions necessary to prevent the spread of serious infectious ailments more faithfully observed. As this amelioration of sanitary conditions will affect mainly the poor, lessening their suffering and adding to their possibilities of happiness, its accomplishment becomes a great Christian duty, obligatory on all those who are interested in the uplifting of the poorer classes.

Professor Virchow, the distinguished German pathologist, used to say that popular medicine was in all ages at least fifty years behind scientific medicine. He had himself discovered the principles of cellular pathology nearly half a century before his death, yet he declared that the popular mind still believed in the old doctrines of humoral pathology,—that is, that the conditions of health and disease depended on the constitution of the fluids of the body (the blood, the bile, the mucus, and so forth), and had not generally accepted modern advances in medical knowledge of the underlying basis of disease in the solid tissues. There is no doubt that many old-fashioned notions long since discredited by physicians are still very generally accepted by the popular mind, and even the intelligent classes sometimes harbour convictions with regard to the good or evil effects of habits of life, diet, and the operation of drugs of various kinds that are entirely contrary to present-day medical knowledge.

It is extremely important, then, that the clergyman or charitable visitor, in giving views on medical matters, which are sure to have much more weight than he perhaps attributes to them himself, should be careful not to make statements for which he has not good authority in modern medical science. It is very easy, in a matter of this kind, to state principles that are not the result of education, properly so called, but are gleaned from early false impressions obtained one knows not how or where, entirely without definite consciousness as to their real origin. The physician himself finds that he is compelled to be careful of this same tendency to put too much stress on traditions with regard to health which he imbibed before he began to study medicine. It is perhaps not so surprising, then, to hear physicians complain often that clergymen instead of being a help are sometimes a hindrance to the enforcement of modern hygienic rules, because they still cling to old-fogy notions of hygiene and sanitation retained from a defective early training. Owing to the influence that the clergyman is sure to exert, this becomes an extremely important matter. Great harm may be done and the physician discredited, almost without a realisation, on the part of the clergyman, that he is interfering in another's department. Sympathetic coordination of clerical and medical efforts would accomplish much good that is now unfortunately left undone.

There is no doubt that for the important crusade against tuberculosis, for instance, the aid of the clergyman will accomplish much for the reduction of the death rate from this disease. What is needed at the present moment is a universal conviction that tuberculosis is not an hereditary but a communicable disease. This does not mean that it is virulently contagious and that as a result sufferers from tuberculosis must at once be segregated from other members of the family and from the community generally; but it does mean that careful precautions must be taken with regard to the disposal of sputum, with the enforcement of the most exacting cleanliness on the part of consumptives themselves. It also means that the person suffering from the disease should not sleep with those as yet unaffected, nor be allowed to live in very close contact, especially with children or susceptible individuals.

The persuasion that tuberculosis is not hereditary will do much to encourage patients suffering from the disease to feel that they are not hopelessly doomed. At the present time it is not unusual to find patients so discouraged, when told that they have tuberculosis, that it is almost impossible to secure a favourable reaction to any mode of treatment. They have seen members of families die one after another, or they have heard stories of the inevitable way in which consumption wiped families out of existence, and they give up hope and become quite cast down. Needless to say, while in this condition any treatment is practically hopeless. On the other hand, the conviction that tuberculosis is only an infectious disease, quite curable in the majority of cases if taken in time, is of itself a most important aid in the treatment of the disease, since courage and faith are the principal requirements for successfully combating the affection.

We have had any number of newly invented remedies for consumption in the last twenty-five years. Scarcely a year has passed in which some new form of treatment, often eventually proved to be the resuggestion of an old therapeutic method, has not been heralded as a positive cure for consumption. In every case the first patients treated by the discoverer of the new remedy have rapidly improved under his care. In the hands of others, however, such results have not been obtained, or only for a very short time at the beginning of the treatment. After a time the new remedy failed in its inventor's hands. The true reason for the improvement was then seen to be, not the remedy suggested, but the favourable influence on the mind of consumptives produced by the faith of the inventor in his remedy, and their reaction to this powerful suggestion when they were put under proper conditions of an abundance of fresh air and a plentiful diet.

This shows, too, the reasonableness of the modern treatment of consumption, which consists not in the giving of drugs, but in securing for the patient a plenty of fresh air for many hours a day and the encouragement to consume a liberal amount of nutritious food. Most of the much advertised remedies for consumption are really harmful rather than beneficent. Many of them are ordinary cough mixtures containing considerable opium, which lessens the cough, it is true, but also lessens the appetite and locks up the bowels. Besides, the cough is nature's method of removing material from the lungs which has become disintegrated, and if allowed to remain will certainly bring about the spread of the infection in the pulmonary tissues. Cough is a natural protective reaction to be encouraged, and is not in itself a source of evil needing to be suppressed. If cough is bothering the patient so much at night as to cause loss of sleep, then it is necessary to make a choice between two evils and somewhat to suppress the cough, even though it involves certain other inconvenience to the patient. All these so-called consumption cures contain materials that are almost sure to disturb the appetite and upset the stomach. The fate of a consumptive patient absolutely depends on his stomach; just as little, then, of medicine must be employed as possible. This will indicate the necessity for clergymen rather advising against than in favour of these proprietary medicines which have been definitely known to do so much harm in recent years. Many a patient delays an appeal to medical aid so long, as the result of trusting to such medicines, that a curable case of consumption becomes incurable, or else develops to such a condition as to require years of treatment on the fresh-air, abundant-food plan, where months would have sufficed before.

A very interesting phase of social medicine is the ease and confidence displayed by people, often of more than ordinary intelligence, in recommending various proprietary medicines of which they know nothing except the fact that someone says he, or more often she, was cured of something or other by their use. A chance remark like this to a sufferer becomes a high recommendation. The hardest problem the doctor has before him is to find out what is really the matter with his patients. Not infrequently people having apparently the same set of symptoms are suffering from quite different ailments. A symptom like a sore throat, for instance, may very well be due to any one of at least a half-dozen of causes, most of which require their own peculiar treatment. When the affection under consideration is as indefinite as a tired feeling, or indigestion, or some one of the many ailments included under the term biliousness or kidney trouble, from which people are supposed to suffer, then the diagnosis problem becomes by far the most serious question in the case, and is often very difficult. The trained physician prudently hesitates, but the inexpert in medicine steps in and quite volubly announces what the ailment is in his opinion, and what will probably do it good. A little knowledge is indeed a dangerous thing in medical matters. If it be remembered that there is a very general impression among medical men now, as the result of recent acquisitions of scientific information with regard to the origin, pathological basis, and course of disease, that very probably more harm than good has been done by the administration of medicines in the past, not only the futility of lay (or clerical) prescribing will be manifest, but also somewhat of the amount of harm that may be done.

It is often a matter for painful surprise, then, to find that clergymen and members of religious communities allow their names to be used in the recommendation of remedies of whose composition they know nothing, for a disease of which they know less, if possible. This evil becomes especially poignant when the columns of our reputable religious press are allowed to be used for the purpose of exploiting the public in these matters. The remedies most often recommended are the so-called tonics. These are best represented by the sarsaparillas, and by various cures for catarrh, indigestion, and kindred indefinite ills, of which there are a great many on the market. These are not secret remedies, since their composition is well known by those of the medical profession who care to secure the information. Some six years ago an analysis of most of them was made by the Massachusetts State Board of Health. [Footnote 1]

[Footnote 1: 28th Annual Report Mass. Board of Health; food and drug inspection, 1897.]

The principal active agent in all of these remedies was found to be alcohol. In most of them it exists in a proportion about equal to that in which it is supposed to occur in ordinary whiskey. Some of them are even stronger in alcoholic contents than the whiskey usually sold in our large cities. This matter has seemed so important that we give the official figures of the Board of Health.

TABLE

From the Report of the Massachusetts Board of Health

Tonics and Bitters

The following were examined for the purpose of ascertaining the percentage of alcohol in each. Some of them have been recommended as temperance drinks!

Per cent, of Alcohol (by volume).

"Best" Tonic 7.6

Carter's Physical Extract 22.0

Hooker's Wigwam Tonic 20.7

Hoofland's German Tonic 29.3

Hop Tonic 7.0

Howe's Arabian Tonic, "not a rum drink" 13.2

Jackson's Golden Seal Tonic 19.6

Liebig Company's Coca Beef Tonic 23.2

Mensman's Peptonized Beef Tonic 16.5

Parker's Tonic, "purely vegetable," "recommended for inebriates" 41.6

Schenck's Sea Weed Tonic, "entirely harmless" 19.5

Atwood's Quinine Tonic Bitters 29.2

L. T. Atwood's Jaundice Bitters 22.3

Moses Atwood's Jaundice Bitters 17. 1

Baxter's Mandrake Bitters 16.5

Boker's Stomach Bitters 42.6

Brown's Iron Bitters 19.7

Burdock Blood Bitters 25.2

Carter's Scotch Bitters 17.6

Colton's Bitters 27.1

Copp's White Mountain Bitters, "not an alcoholic beverage" 6.0

Drake's Plantation Bitters 33.2

Flint's Quaker Bitters 21.4

Goodhue's Bitters 16.1

Greene's Nervura 17.2

Hartshorn's Bitters 22.2

Hoofland's German Bitters, "entirely vegetable and free from alcoholic stimulant" 25.6

Hop Bitters 12.0

Hostetter's Stomach Bitters 44.3

Kaufmann's Sulphur Bitters, "contains no alcohol." As a matter of fact, it contains 20.5 per cent, of alcohol and no sulphur 20.5

Kingsley's Iron Tonic 14.9

Langley's Bitters 18.1

Liverpool's Mexican Tonic Bitters 22.4

Paine's Celery Compound 21.0
Pierce's Indian Restorative Bitters 6.1
Puritana 22.0
Porter's Stomach Bitters 27.9
Pulmonine 16.0
Rush's Bitters 35.0
Richardson's Concentrated Sherry Wine Bitters 47.5
Secor's Cinchona Bitters 13.1
Shonyo's German Bitters 21.5
Job Sweet's Strengthening Bitters 29.0
Thurston's Old Continental Bitters 11.4
Walker's Vinegar Bitters, "contains no spirit" 6.1
Warner's Safe Tonic Bitters 35.7
Warren's Bilious Bitters 21.5
Wheeler's Tonic Sherry Wine Bitters 18.8
Wheat Bitters 13.6
Faith Whitcomb's Nerve Bitters 20.3
Dr. Williams' Vegetable Jaundice Bitters 18.5
Whiskol, "a non-intoxicating stimulant, whiskey without its sting" 28.2
Colden's Liquid Beef Tonic, "recommended for treatment of the alcoholic habit" 26.5
Ayer's Sarsaparilla 26.2
Thayer's Compound Extract of Sarsaparilla 21.5
Hood's Sarsaparilla 18.8
Allen's Sarsaparilla 13.5
Dana's Sarsaparilla 13.5
Brown's Sarsaparilla 13.5
Corbett's Shaker Sarsaparilla 8.8
Radway's Resolvent 7.9

The dose recommended upon the labels of the foregoing preparations varies from a teaspoonful to a wineglassful, and the frequency also varies from one to four times a day, "increased as needed."

Many so-called tonics not on this list are also known to contain alcohol, though not as yet officially analysed so as to give exact figures. Most of the cure-alls for women's ills contain alcohol in noteworthy amounts, this being in fact usually the only active ingredient in them.

As the analyst of the State Board of Health of Massachusetts is a thoroughly competent chemist, and as these figures have now been before the public for over five years without any contradiction on the part of the manufacturers of these remedies, though it is evident how undesirable the truth of the matter is from an advertising standpoint, there can no longer be any question as to the authoritativeness of the proportions of the alcohol in the remedies as given.

It is rather sad to think of mothers giving these remedies to their children, hopeful of the good they may accomplish, when, as a matter of fact, it would be so much simpler and just the same in the end, to give them, instead of a tablespoonful of the favourite sarsaparilla, whatever it might be, a tablespoonful of dilute whiskey. As was noted in the volumes on the Physiological Aspects of the Liquor Problem published recently by a sub-committee of the Committee of Fifty for the investigation of the liquor problem, not a few prominent total abstinence advocates have put themselves on record as recommending these remedies, though there can be no possible doubt of the great harm likely to arise from their use. There are many physicians who feel sure that some of the alcoholic habits in women, whose origin it has been hard to account for, were really contracted during this secret "tippling" process under the form of a tonic remedy. Everyone knows that any tonic, in order to be effective, has to be gradually increased, so it is not surprising that in many cases physicians have heard of patients taking six to ten tablespoonfuls of some tonic remedy every day. This would be the equivalent, in some cases, of from three to five ounces of whiskey—a rather liberal allowance even for a confirmed whiskey drinker.

As noted by the Massachusetts Board of Health, the dose recommended upon the labels varies considerably, but practically all agree in suggesting that the amount of the remedy taken shall be increased as needed. A simple presentation of this subject will surely be sufficient to arouse clergymen to a proper sense of their duty in this matter. Senators, judges of Supreme Courts, Congressmen, and even university professors and teachers may be so benefited by dilute whiskey, taken early and often, as to be tempted to furnish testimonials for them (for a due consideration usually), but clergymen should at least know something of the consequences of their act before committing themselves.

An almost precisely similar state of affairs obtains with regard to another class of favourite popular remedies. A number of so-called blood-purifying remedies have been recommended at various times, and here, as in other things, it is surprising to find how many intelligent people lend themselves to the exploitation of the public in the interests of the proprietary vender, who cares only to sell, and cares very little what effect his remedies may produce. Most of the sarsaparillas are said to be blood purifiers. It is surprising what vogue this word "sarsaparilla" has obtained. A little more than half a century ago a German chemist and pharmacist announced that the sarsaparilla plant contained certain principles that could be extracted by boiling, and that form excellent remedies for atonic and anaemic conditions. This announcement was received by the medical profession very kindly, and immediate tests as to the efficacy of the new remedy were made. As a result of these tests, within a few years the inefficacy of sarsaparilla became very clear. It is almost entirely without effect upon the human system. In the meantime, however, the word "sarsaparilla" was one to conjure with for the popular mind, and the sarsaparilla remedies began to be manufactured. Millions have been made on them and out of the public. The only active agent as regards tonic qualities which they contain is, as we have said, alcohol. Most of them however, contain at least one other well-known drug likely to be at least as harmful as alcohol. This is iodide of potash. Very few of the so-called sarsaparillas are without a notable proportion of this strong mineral salt, as the Massachusetts Board of Health said.

"With but few exceptions they contain a considerable percentage of a very active and powerful remedy, the iodide of potassium. The sale of such an article in unlimited quantities by druggists, grocers, and others is censurable. More than this, the method of its sale is dishonest, since the unwary purchaser is led to believe that he is purchasing a harmless vegetable remedy, namely, sarsaparilla.

"It may be seriously questioned whether the blood of persons who take iodide of potassium continuously is not decidedly impoverished, instead of being purified, as is claimed by the manufacturers. It is not uncommon to find persons who have used continuously six, eight, or ten pint bottles of one of these preparations.

"Unlike sarsaparilla, the iodide of potassium is classed among poisons by nearly every writer upon toxicology."

Practically all the proprietary remedies have their most potent principle in the supposed mystery of their composition. As a matter of fact, all are simple prescriptions, well known to physicians, and owing their successful treatment of many ills much more to the printer's ink used to secure their sale than to any pharmaceutical ingredient which they contain. No important remedy has ever been put on the market by advertising methods. Exposure of the charlatanry of such methods will not, however, cause an interruption of their sale. Long ago Barnum said that people wish to be humbugged, and there is no doubt that they have been, are, and will be humbugged just to the extent to which they lay themselves open to the alluring methods of the advertiser. It does seem too bad, however, that the influence of the clergymen and of religious as well as charitable visitors—an influence acquired because of the confidential position they occupy and the feeling of good faith their mode of life inspires—should be abused for the encouragement and extension of what is manifestly a great evil.

Alcohol and iodide of potash are not the only drugs likely to do harm that are incorporated in proprietary medicines. Great complaints have recently been made with regard to the spread of the cocaine habit in this country. Not a few of the remedies that are supposed to give immediate relief to colds in the head contain cocaine in dangerous amounts; and there seems no doubt that in many cases the drug habit for this substance has been acquired innocently and unconsciously at first by the use of such preparations. These are only the more notable evils likely to result from the indiscriminate employment of medicines of whose composition there is complete ignorance, and of whose effect there can be only the judgment dependent upon the subjective feelings of the patient. It must not be forgotten that the patient's feelings are for the moment often favourably influenced by some substance that may do no good to the ailment, though making the patient less sensitive to any symptoms from which he was suffering; but in the end doing positive harm, because of the contraction of the alcohol or some drug habit, or because the suppression of symptoms may be the very worst thing for the patients, since it allows the underlying ailment to progress to a serious stage without forcing them to have it treated in radice.

These are only a few examples that show very well the inadvisability of recommending in any way medicines of which one does not know the exact contents. The present writer has had one example of how utterly disingenuous, though one feels much more like calling it rascally, the manufacturers of so-called patent medicines or proprietary remedies may be. One of the remedies widely advertised for the cure of epilepsy, or fits, is announced always as containing no harmful drugs, no bromide of potash. The manufacturer of the remedy was asked how he could say any such thing, since it was very evident even to the taste that the remedy contained bromides. "Oh," he said, "yes, it contains sodium bromide, but not bromide of potash." Almost needless to say, sodium bromide is at least as harmful as potassium bromide, and the advertisement is entirely for purposes of deception.

The poor epileptics have been a source of revenue for quacks and charlatans as long as history runs. At the present time one not infrequently finds testimonials from convents, asylums, reformatories, and the like, asserting the value of some particularly advertised remedy for this disease. All these remedies contain bromides. The treatment of epilepsy is now better understood by physicians and it is generally recognised that the two things that epileptic patients need are outdoor air and as far as possible all freedom from responsibility. Bromides will, for a time, control the number and frequency of the attacks, but if used indiscriminately, and especially if employed without any proper realisation of their possibilities for harm, these salts are almost sure to make the condition of the patient much worse than before, to bring on a state in which mental symptoms predominate over physical, and in which the patient may go into dementia, or some form of mental alienation. Especially is this true with regard to epileptic children. Continuous dosing with drugs of any kind is sure to do them harm rather than good. Care for their diet and rest and the removal of all sources of disturbance of their digestive tract is more important than any other method of treatment.

The poor children have to suffer many things from many people. People hesitate, as a rule, to accept recommendations with regard to the administration of drugs to their animals when the person who gives the recommendation is known not to be an expert in the matter. Almost any suggestion, however, with regard to the dosing of their children is likely to be followed by loving but indiscreet mothers. It is well known now, and in many cases is admitted, that the so-called soothing syrups so often given to children contain opium in quite appreciable quantities. Needless to say, nothing much worse than this could possibly be given to children. The child soon becomes accustomed to its daily dose of opium and craves the repetition of it. It will not sleep without it, and as this adds to the sales of the remedy, this special ingredient continues to put money in the pockets of the manufacturers, but at the expense of the nervous stability of the child, and lack of resisting power later in life. It would be hard to say how many of the nervous wrecks so commonly met with in young adults now are to be attributed to this unfortunate state of affairs early in life; but undoubtedly this evil has had much to do with the noticeable increase in the nervousness of our people. The more nervous the heredity of the child, the more it must be guarded against such mistaken methods of inducing sleep, or the result is sure to be serious.

Scarcely too much can be said in condemnation of most of the proprietary remedies for constipation, though it is in this department of medication that the non-medical are freest with their advice. First, the cheapest possible drugs are selected by the manufacturers of such remedies. Secondly, those drugs especially are employed which, while producing the desired immediate effect, are always followed by a reaction which requires further use of the medicine. One finds testimonials, however, from all classes of the community, even from clergymen, with regard to such remedies, though at the last international medical congress it was confidently asserted, by three of the most prominent specialists in digestive diseases in the world, that the modern problems in digestive disturbances are so much more intricate than they used to be, and the affections which develop are so much more difficult of treatment, because of the use of these unsuitable remedies, and the consequent habituation to drugs, which has been acquired during the prolonged period of their employment.

In recent years catarrh has become the word that is supposed to attract popular attention most, and accordingly is the watchword of the proprietary medicine manufacturer. A long time ago, that is, about half a century ago, catarrh was supposed really to mean something in medicine. Those were the days of humoral pathology, when disturbances of secretion were supposed to be the basis of all disease. Accordingly, whenever there was an excessive discharge from the nose, a patient was said to be suffering from catarrh, and as the nasal secretion was supposed to be connected in some way with the brain, it is easy to understand how significant such a pathological condition might well be thought. In more recent years, the word "catarrh" has still been employed by physicians who thoughtlessly employ terms that they think will be better understood by the laity, owing to their familiarity with them, though they have been outlived in medicine. From representing an affection of the nose, catarrh, as a consequence, has come to be employed for an excess of secretion from any mucous membrane. Accordingly we hear of catarrh of the stomach, catarrh of the bladder, or catarrh of the bile-ducts, and there has come to the general public a notion that catarrh is an all-pervading affection whose ravages must be prevented, at all hazards, and whose beginning must be the signal for prompt medical treatment.

As a matter of fact, catarrh, when it means anything, means only that stage of inflammation in which there is an increased secretion and which represents an inflammatory condition so mild as often to be described as only hyperaemic, that is, due to an increase of blood in the part. It is rather easy to understand that if more blood flows through a mucous membrane, there will be greater secretion from it than would normally be the case. This is what happens in the production of catarrh. As a rule, it is only a passing congestion without any lasting changes in the tissue. Catarrh may, however, continue to be present if the irritation, which originally caused the congestion, be allowed to continue. It is this irritation, however, which needs to be treated, and not the catarrhal inflammation, which is only a symptom of it. The three most used words in popular medicine,—catarrh, rheumatism, and gout,—when traced to their etymological signification, mean the same thing. Catarrh means a flowing down, rheumatism a state of flowing, both being formed from the Greek verb , to flow, while gout is derived from the Latin word gutta, a drop, which hints at the excess of fluid that is supposed to be the basis of the disease.

For these three diseases, however, the most varied remedies have been proposed, and practically entirely without success, when tested, in a large number of cases. As a matter of fact, under the two words catarrh and rheumatism, there is grouped a series of affections very different from one another, and requiring very different treatment. The important thing is not so much the suggestion of a remedy as the recognition of the particular cause which in one case is producing an excess of secretion and in the other is giving rise to the so-called rheumatic pain. When the exact cause can be found, it is usually not so difficult to succeed in preventing the recurrence of the troublesome symptoms. It is with regard to these two diseases, however, that in non-medical circles even intelligent men are ready to give advice. They constitute the most puzzling problem that the physician has to deal with, but the non-medical mind waives the difficulty and suggests the remedy. In this matter one is forcibly reminded of a famous expression of Josh Billings, who used to say, "It is not so much the ignorance of mankind that makes them ridiculous as the knowing so many things that are not so."

Clergymen, lawyers, members of Congress, and of various state legislatures, all permit their portraits to appear, advertising the merits of some trumped-up cure for catarrh or rheumatism. It is interesting to realise, then, that in most cases, according to expert testimony, the remedy they recommend so highly consists of nothing more than diluted alcohol flavoured so as to taste like medicine. The only real effect is the alcoholic exhilaration which follows its ingestion and gives the sense of well being, because of which the testimonials are provided. As one of the medical journals said recently, it would be very interesting to make a list of the men and women throughout the country who, by permitting their portraits and recommendations to be used in the advertisements of various patent medicines, have practically confessed that they like to take their whiskey rather dilute but mixed with a little bitters. The whole question illustrates the tendency of the proprietary medicine man to exploit some phase of medicine long after it has ceased to be of interest to the medical profession.

With regard to all of these things clergymen may do a great humanitarian work by protecting the poor from the efforts of advertising remedy-makers to get their hard-earned money. It is sometimes said that long years have been spent in the preparation of a remedy. This not only is never true, but never has been true in the history of proprietary medicines. Some one who has an eye to business gets hold of a prescription of which he knows nothing, but of which his advertising agents are able to say much, and the result is sometimes a fortune for the advertiser. There is always a pretence of philanthropy, but it is the mask of heartless hypocrisy. Unfortunately many of our religious journals are tempted by the promptly paid bills of such manufacturing concerns to print their advertisements. They are aiding in a deliberate swindle, and if this were better understood there would be much less suffering and fewer vain hopes. The best-managed newspapers and magazines in the country are now absolutely refusing all medical advertisements. This is the only proper attitude in the matter, for there is a place to advertise medicines, if they are worthy, and that place is the medical journals. If the popular advertising could be reduced, we should soon have much less of the proprietary medicine evil.

There are many ways in which clergymen by their example, their advice, and their influence can be of great assistance to practitioners of medicine. It is very sad, then, to find that some of them, having elaborated theories of their own on certain subjects, or having taken up with peculiar notions, are in opposition to the accepted medical teaching of the world. Occasionally they are found among the ranks of the anti-vaccinationists, though if there is anything that has been demonstrated to a certainty, it is that vaccination has practically eradicated smallpox, considering the frequency of the disease a century ago, and that it would absolutely eradicate it, if the practice could be made universal. Statistics are at hand to demonstrate this beyond all possibility of doubt. There are a certain number of people, however, who apparently, out of a desire for singularity as much as anything else, refuse to accept the evidence. It is very unfortunate to find clergymen among them, for it tends to bring the clerical judgment into disrepute.

Nearly the same thing might be said of antitoxin for diphtheria. Clergymen seem to consider it necessary for them to have their minds made up as to whether the use of diphtheria antitoxin is advisable or not. If they have once committed themselves to the expression of the opinion that antitoxin is of no value, then no amount of evidence will succeed in changing their opinion. Under these circumstances it becomes extremely difficult at times for physicians to succeed in having families permit them to treat their patients after the manner in which they are convinced the treatment should be carried on. If such clergymen would only realise that the clergyman has, as a rule, much less right to express opinions on medical subjects than has the physician to air views with regard to theological principles, there would be much less friction, and it would be better for patients in the end.

There are certain sanitary regulations that clergymen should not only not oppose, but endeavour, by every means in their power, to have those who respect their opinions follow out as carefully as possible. Such sanitary regulations have in the past twenty-five years practically cut down the death rate of our large cities a half. There is no greater source of alleviation for the physical evils, at least those which afflict the lower classes, than the due enforcement of modern sanitation. There are prejudices, however, that must be overcome, and the clergyman should be found beside the doctor, helping him rather than opposing him, as is sometimes the case.

JAMES J. WALSH.

SOME ASPECTS OF INTOXICATION

There are various drugs that, through acute or chronic poisoning from their use, cause mental disturbance,—alcohol, chloral, cannabis indica, somnal, sulphonal, paraldehyde, ether, chloroform, antipyrin, phenacetin, trional, chloralamid, iodoform, atropine, hyoscyamus, salicylic acid, quinine, lead, arsenic, mercury, opium and morphine, the bromides, cocaine, and others. Of these intoxicants alcohol always has been most commonly used by western nations, but the moral aspects of alcoholism have not been shown with sufficient insistence. There are many sots in human society much less reprehensible than to the unskilled observer they appear to be; others are more blameworthy.

Morality, as far as the agent is concerned, apart from the nature and circumstances of the deed, supposes, first, voluntary acts, or acts that proceed from the will with a knowledge of the end toward which the acts tend; and, secondly, free acts, or acts that under given conditions may or may not be willed. If by unavoidable chance one stumbles against a man standing at the edge of a wharf, knocks him into the water, and drowns him, the act has no element of morality in it, because it is not voluntary and free. If a mind is diseased, and, impelled by a mad notion of persecution, it brings about a like killing, there is no question of morality, because the agent is not free, and when fully analysed his action is not voluntary.

An act is more or less voluntary and free, and therefore more or less moral, as the agent is affected by ignorance, passionate desire, fear, or disease. Ignorance, fear, and disease may be such as to remove all quality of morality from an act. Certain diseases or pathological conditions, especially of the nervous system, can take out of an act the elements of voluntariness and freedom that are necessary to make the act moral or immoral, provided, however, these pathological conditions are not brought on through the fault of the subject in which they exist. If a man voluntarily becomes drunk with alcohol, or some other drug, he is, of course, accountable for the evil he may unconsciously do while under the influence of that drug, and if he begets an idiot or a criminal imbecile in his drunkenness, he must atone somewhere for the blinded soul of his child. Here, again, there are certain extenuating circumstances, because very few drunkards are fully conscious of the extent of the evil in alcoholism.

Apart from the other requirements that go to make an act moral, the agent must be sane; that the act be immoral, he must be sane or insane, either temporarily or permanently, through his own fault; that it be devoid of morality an act must be a mere actus hominis, or it must be the act of a person blamelessly insane. If a man knows that an alcoholic is liable to beget a criminal imbecile solely because of the alcoholism,—and most men are aware of that fact,—this father or grandfather is more or less accountable for every larceny, rape, and murder done by the imbecile. The law, therefore, should put the imbecile into safe keeping, then seek out the father and hang him.

Insanity is a common condition, but it has not been satisfactorily defined. It supposes an appreciable unsoundness of the will, memory, and understanding, or of one or two of these faculties, but no alienist has given a short differentiation of that unsoundness. Where shall we draw the line between the weak but responsible will and the insane will? What degree of opacity between intellect and the world separates the ignorant man from the lunatic? The extremes of sanity and insanity are readily recognisable, but the intermediate degrees are not clear. There is no test to apply to all cases; each must be diagnosed from its peculiar symptoms, but the will of an insane man is always weak. It can not deny or defer the gratification of a desire, nor can it keep up an effort. Even in its lightest forms insanity is selfish and impolite, because it lacks the force of will necessary to take trouble. It foregoes great future benefit for slight present gratification. The insane man is idle, or busy only in work that he likes, in pleasurable activity. A marked quality of sanity is the capacity for sustained work, and the man that shirks work merely because he does not like it is gratifying himself dangerously.

These defects are found commonly in sane persons, but the lunatic can not rise from them, and he adds to the defects of will a warped intellect. He can not adjust himself to his surroundings, and the fault is in himself, not in the circumstances. His intellect may be brilliant, but it sooner or later shows a taint. The insane man is not a free, rational agent.

Alcoholism readily passes over into unmistakable insanity, and it almost always is the cause of nervous degeneration in the children born within its influence. This, is a phase of the evil not sufficiently insisted upon by those that plead for total abstinence.

Chronic poisoning by alcohol induces hardening and calcification in the walls of the arteries, degeneration of the nerve cells and dendrites, wasting or overgrowth of the heart muscle, and fatty degeneration of the liver and kidneys. The nerve centres that control the circulation of the blood are paralysed by it, and, as a sequence, the arteries and capillaries are diminished in calibre. This state in turn obstructs the flow of the blood, and the body is not nourished, nor are the waste and poisonous results of metabolism carried off as they should be. Alcohol prevents the haemoglobin of the blood from doing its office, which is to supply oxygen and remove carbon dioxid. It absorbs the necessary water from the tissues, and thus it acts as a corroding poison. It is also a functional toxin, because it depresses the activity of organs by injuring the innervation. The poison affects the brain, and as the cerebral gray matter, especially its pyramidal cells, are the physical instruments of thought, will, and memory, or the means of communication between the soul and the outer world, the exercise of these spiritual functions is checked or inhibited by it.

A tendency to excess in the use of alcohol commonly manifests itself before the thirtieth year, and in some cases it may be removed at the alcoholic climacteric, which is from the fortieth to the sixty-fifth year. Those that become drunkards are usually of a neuropathic constitution, through inheritance or abuse. Severe diseases, like influenza, syphilis, typhoid; injuries to the head, sunstroke, shock, worry; the disturbance that may accompany puberty, pregnancy, lactation, and so on,—cause a nervous depression which is soothed by alcohol, and thus a habit is fixed. The reckless prescription of alcohol by some physicians is another cause of the habit, and the use of proprietary medicines is a still more prolific source of drunkenness and the consequent misfortune.

Cider, beer, ale, and porter contain from 4 to 6 per centum of real alcohol; light wines, red and white, and natural sherry, 10 to 12 per centum; strong sherry and port, 16 to 18 per centum; brandy, 39 to 47 per centum by weight, or 46 to 55 per centum by volume; and whiskey, 44 to 50 per centum by weight, or 50 to 58 per centum by volume. The effect of these liquors on the body is due primarily to alcohol, and secondarily to ethereal derivatives of alcohol. Some owe a part of their effect to non-volatile substances,—beer from which all alcohol has been boiled can still affect the body in a marked degree.

The chemist of the Massachusetts State Board of Health (Document No. 34) gives the percentage of alcohol in the common proprietary medicines, and these percentages will be found in the article on Social Medicine.

The weakest of these compounds are twice as strong in alcohol as beer, and they treacherously bring about the habit of drunkenness in disposed persons who may be very desirous to avoid such a calamity.

Some men and women are quickly destroyed by alcohol; others resist it more or less successfully for a lifetime, as far as mere existence is concerned. Alcoholism is one of the commonest causes of insanity, but it is often an effect of insanity. It may be an early symptom of paresis, or a part of the maniacal stage of circular insanity. In poisoning by alcohol the higher nerve centres are first affected and the lowest last. The sense of human dignity and of morality, the exercise of the intellect, are more or less inhibited before the motive muscles are affected.

The usual effect of alcoholic poisoning is boisterous exaltation of mind, but there is a depressed type of drunkenness which weeps. Some patients at once are subjected by hallucinations and delusions, others are so depressed that they have a suicidal tendency, others may have a maniacal frenzy that is destructive or homicidal. In these neuropathic conditions muscular co-ordination is commonly well preserved—the patient is "drunk in the head and sober in the legs."

In alcoholism the mental changes are gradual and progressive. The intellect is blunted, the judgment becomes foolish, the moral sense is dulled. The drunkard is always a liar. Delusions not infrequently occur, and it is one of the common symptoms of alcoholic insanity to suspect a wife or husband of conjugal infidelity. If a man that is a drunkard accuses his wife of infidelity, the chances are fifty to one that she is innocent and that he is in the first stages of insanity. This symptom is characteristic also of cocaine intoxication.

Another mental disturbance of acute alcoholism is delirium tremens, which is inexactly called mania a potu by some writers. Delirium tremens is not a form of mania, but an acute hallucinatory confusion, in which the consciousness is much more impaired than it is in a mania. Mania a potu is a real mania, and it is transient commonly, although it may leave permanent mental weakness with delusions.

In chronic alcoholism a paranoid condition may occur, and this often is incurable. This psychosis may come on suddenly or gradually. In true paranoia the delusions are systematised, but in this alcoholic pseudoparanoia the enfeebled intellect can not build up coherently even a delusion. The alcoholic hallucinations are visual and auditory, and we find delusions of persecution, especially of a sexual nature. The patient hears all kinds of insulting remarks made by "voices." These voices often come from his own belly. His enemies send poisonous or foul odours into his room at night, and the groundless suspicions of his wife's infidelity take most outrageous forms of expression. He will swear he has seen her misdeeds. Often the baseless suspicions of his wife begin before any other noticeable impairment of intellect, and are not recognised as delusions. The first step a priest should take in investigating accusations of conjugal infidelity is to find out whether the accuser is a tippler or not.

The delusions of persecution lead to attacks on the supposed enemies which often are homicidal. Occasionally alcoholic insanity takes on a paretic form, or it may be epileptic. Ten per centum of alcoholics are epileptic. When the children of alcoholics are epileptic, the convulsions begin in these children about four years earlier than in children that are epileptics from other causes. If epilepsy is latent, alcoholism will start it into action.

Alcoholism sometimes produces a condition of waking trance followed by amnesia (lack of memory). In such a state the drunkard may transfer property, carry out complicated professional actions, commit crime, take long journeys, travel for days, and so act that no one notices his disordered mental condition. Then suddenly he awakens and he has no recollection whatever of what has happened during this trance. He appears to be conscious, but to have no memory of his consciousness. There is another alcoholic amnesia, found especially in those that drink much during the morning hours, where there is instantaneous forgetfulness. If you ask one of these men to shut a door, for example, he will forget between his chair and the door what he started to do. This condition is difficult to cure even after the use of alcohol has been relinquished.

Dipsomania is a form of impulsive degenerative insanity, and it is probably epileptic in origin. After a few days of insomnia and loss of appetite for food, there is an irresistible impulse to drink alcoholic liquor and to indulge in other excess. The patient drinks until all means of getting alcohol are exhausted. He will take crude alcohol, bay rum, cologne, the alcohol that is about pathologic specimens in a hospital museum. The attack lasts from one to two weeks, and is followed by depression and a feeling of remorse. The onsets are irregular in occurrence, and between them the patient may be temperate or have even an extreme distaste for alcohol. This form of disease is not infrequent among professional men and clergymen, and it is impossible to find out just how far the patient is responsible for his condition. If bishops would investigate the alcoholic tendencies of the families of candidates for seminaries, and reject all that have this taint, there would be much less scandal. It is a serious error of judgment to ordain a seminarian that has even once been under the influence of alcohol, and those seminarians that cover up the tippling of a companion, because he is a good fellow, are guilty of far-reaching crime. The fact is worth investigation whether or not a liquor dealer who never drinks alcohol, but who lives for years in the presence of volatilised alcohol, has much of the alcoholic degeneracy and a tendency to beget neurotic children. Certainly the fumes of wood alcohol have killed workmen that went down only once into a vat containing these fumes, and other alcohols in the form of vapour should have deleterious effects. Féré produced monsters in chickens by exposing eggs to the vapour of alcohol.

In judging a drunkard, it must be remembered that in many forms of alcoholism, after the condition is well established, the patient has little more freedom of will than a brute has. If he is accountable for the habit, he is blamable for the crime that follows. If he is not accountable, and it is often very difficult to prove that he is, he is to be treated as a blamelessly insane man. In proper surroundings, and with skilful direction, a child born with a tendency, or more exactly a temptation, to dipsomania or other alcoholic neurosis can be saved, but commonly the circumstances of such a child's life are the worst imaginable. These children must never take alcohol, even as a medicine, and they must not be pushed in school to nervous exhaustion. A tendency to unchastity can "run in families," like a disposition toward alcoholism, but the disgrace in yielding to this vicious bias keeps many such unfortunates clean. It is to be regretted that public opinion can not give the same aid in alcoholic predisposition.

A confirmed alcoholic should be prevented, if possible, from marriage, because his sins will be visited upon his posterity. The first children of an alcoholic may be mentally sound, the younger children are more or less mentally weak, the youngest are not uncommonly imbeciles, or idiots, or under shock they grow insane. Fortunately many of the children of alcoholics die at an early age, and the family of a drunkard very seldom lasts beyond four generations. In the first generation moral depravity and alcoholic excess are found; in the second, chronic drunkenness and mania; in the third, melancholia, hypochondria, impulsive and homicidal ideas; in the fourth, idiocy, imbecility, and extinction of the family. The lower the social caste of the drunkard, the greater the liability of meeting these blights.

Priests should take a deep interest in societies established for the promotion of temperance, and the only temperance for most persons is total abstinence. No man knows what latent tendency to alcoholism he may have, especially in America, where great grandfathers are unknown and the climate and life are trying on the nervous system. The adulterated liquors sold everywhere at present make the danger greater than it ever was. Whatever may be the truth as regards heredity, there is no doubt concerning the strong influence of environment; therefore get into the temperance societies the children of alcoholic parents, of parents that are shiftless, hysterical, irritable. If a man has a violent temper or if he is unchaste, get him and his children into the society to check the downward drift. A bad temper is a neurotic taint, and it commonly is a first step toward alcoholism. Do not forget to warn the people against patent medicines that contain alcohol.

If you go over the list of the families in a parish, it is startling to find how few there are without one or more "black sheep." The human black sheep, in a good environment, is always physically imperfect, and never so black as the gossips paint him. He may be a powerful football player, but there is something wrong with his gray matter. He is morally deaf, he was born so, and he is to be excused if he can not always hear the still, small voice. This may sound like lax doctrine, but it is true, nevertheless.

We must recognise that moral weakness is very often, partly at least, a physical defect, and there is no such state as "moral insanity" where the intellect is normal. Now, I do not wish to be quoted as holding that all moral depravity has a physical basis; most of it is the unalloyed stuff; the Lombroso criminal is not a scientific fact; but there is a moral condition very frequently met with which is largely physical in origin. Given so many grains of cocaine or morphine or so many ounces of alcohol, and you can make a liar of a man once on the way toward sanctity. Given an attack of hysteria in a holy nun, and she at once becomes a liar, an altogether blameless liar, but no influence that does not remove the physical cause will cure the lying.

The morally weak do not at present obtain enough religious instruction. Their religion is more a matter of inheritance and habit than of positive energy. It is "in the bones," sometimes in the fists, rather than in the soul. They prefer the Sunday newspaper to the Sunday sermon. The remedy here seems to be in making the Sunday school solidly interesting and its teaching impressive.

Alcoholism in the parents, especially drunkenness at the moment of conception, is one of the chief causes of idiocy in children. Féré, as was said before, by injecting a few drops of alcohol beneath the shell of hens' eggs, or by exposing the eggs to the vapour of alcohol, could produce monsters almost invariably. In 1000 cases of idiocy at the Bicêtre, Bourneville found a history of alcoholism in 620, or 62 per centum: in the fathers of 471, in the mothers of 84, in both parents of 65; and in one-half of the remaining 38 per centum no history was obtainable—probably most of these also had the alcoholic taint. The administration of alcohol to infants, of gin and whiskey, of essences of peppermint and anise, to relieve colic or induce sleep, and the dosing with opiates like paregoric, are also well-established causes of idiocy.

The idiot is practically dead, except for the trouble he gives in caring for him; but another unfortunate, the imbecile, most commonly the offspring of alcoholics, is often capable of great mischief. The higher grades of imbeciles, those nearest the normal, are almost invariably criminals. Not all criminals, of course, are imbeciles, but a vast number of petty and brutal criminals are imbeciles. We keep these unfortunates most of their lives in jail, while we fine their drunken fathers, the cause of the imbecility, "five dollars and costs."

Imbecility has grades,—from marked lack of intellectual power, a stage little beyond idiocy, up to the presence of a mind capable of fair education,—but in all cases there is real defect, either of intellect or of will. Sometimes, where the will is so weak that the patient becomes a criminal in spite of all training, the intellect is practically normal to the superficial observer.

The grades of imbecility can not be clearly marked off from one another, but, roughly speaking, there are three. The lowest grade of imbeciles understands simple commands, and has a slight manual dexterity. They express themselves by signs and in monosyllables. They can not concentrate attention upon anything, nor can they be taught to read or write. Careful training can advance them so far that they may do rough, menial work, and they are industrious when directed by a present superior. They are inclined to masturbation. If they are not teased, they are quiet; if annoyed, they may become dangerous.

Imbeciles of the middle grade can converse in a narrow vocabulary, and they commonly stammer. They may be taught to read monosyllables; they can not do even the simplest sum in addition, yet they show a certain shrewdness. They are irritable and quarrelsome, inclined to lying and stealing, and they have no sense of shame. They will not do any regular work, but change from one occupation to another. They may have sexual instincts and cause trouble on that account. They are slow to understand, their memories are defective, and they are always very vain. Their belly and what they shall wear are the chief things in their lives. They are less criminal than the highest grade of imbeciles.

The third class, the high-grade imbecile, is the most important, because he is commonly a criminal. His intellect is below the average, and his will is very flabby. He learns little at school, and what he does learn is acquired slowly. He reads and writes badly and he may be able to add simple columns of numbers, but he can not multiply or divide. Sometimes such an imbecile has a remarkable facility in getting a speaking knowledge of two or three languages, and he may learn a trade. There is a high-grade imbecile that is cunning and shrewd, but he has no will, and he is a criminal. As imbeciles approach the normal in intellect they recede from it in abnormality of will.

Autopsies on imbeciles show an infantile development of the forebrain. Imprisonment does no good in these cases. They are not taught anything in prison, not even a trade, because the labour organisations and the protected industries will not permit prison labour. They should be confined so that they will not pervert youth and propagate their kind. It is impossible to say how far a given imbecile is morally accountable for what he does, but the accountability is not full in the best cases. A neurasthenic, however, is not to be mistaken for an imbecile. A neurasthenic person may have a tender conscience, an imbecile has no conscience. In imbecility the fault is in the will, rather than in the intellect, in the middle and highest grades. Many women, especially, that are hopeless fools intellectually have strong wills, but an imbecile never has a strong will except in the sense that stubbornness is strength. Stubbornness is perverted strength.

Morphine and Cocaine Intoxication,—Morphine, an alkaloid of opium, is used very extensively as an intoxicant. Since 1890 the importation of opium into the United States has increased fourfold, although physicians are now using less opium than they formerly did.

The insomnia, worry, moral distress, which bring on the alcoholic habit in some persons, lead to morphinism in others. Some physicians, by carelessly prescribing morphine for neuralgia, migraine, dysmenorrhoea, or any pain, make their patients slaves of this drug.

The degenerative effects of morphine are not so great nor so rapid as those of alcohol. It does not shorten life so much as alcohol does, nor are the children of a person addicted to the use of the drug so liable to idiocy and imbecility. The mind is enfeebled—slowly in some cases, rapidly in others. The patient will resort to almost any means to obtain the drug, if he is deprived of it. Authorities hold that he will lie without reason, merely for the perverse pleasure in deceiving, that he is uncertain and treacherous, with a dull conscience and morbid impulses. There are exceptions to this in cases where the drug is easily obtained by the patients. Opium and morphine diminish the sexual appetite in males, even to impotence. The bodily changes are slow but profound.

When a user of morphine has been deprived of the drug for from ten to fifteen hours, he becomes so weak he can not stand; he gets diarrhoea with cramps; he sweats, trembles, and collapses. Later, mental disturbance comes on. He grows delirious, sees insects and small animals, as the delirium tremens patient does, and his suffering is very great. It is extremely difficult, and commonly impossible, to cure the morphine disease after it has been firmly established, and a deliberate acceptance of the habit is evidently a grave vice. Where a patient has become addicted to the use of the drug, through the fault of a physician, or through ignorance, the treatment from the social point of view of such a patient is commonly cruel.

Cocaine intoxication is much worse than morphinism. It is a new excess, which was unknown before 1886. Many users of morphine can carry on business, but the cocaine habitué can not do so. He is always extremely busy doing nothing. He writes long letters which are never finished. He changes from work to work, and even his conversation wanders. His bodily weight decreases rapidly, even one-third of his whole weight may be lost within a few weeks. The skin hangs in folds and is of a dirty yellow colour, the facial appearance is that of extreme distress, and the muscles are feeble. Fainting, irregular cardiac action, sweating, and insomnia are other symptoms.

Insanity is an occasional sequence, with hallucinations, especially of hearing. Such a patient hears roaring noises and voices; his secret thoughts are shouted out, he thinks, to crowds; loud screams, shrieks of murder, and similar noises appall him. Again, he sees swarms of flies, ants, roaches, which cover him and crawl into his mouth, nostrils, and ears. He feels bugs crawling under his skin, and he has a multitude of similar interesting experiences.

Such patients grow homicidal. Like alcoholics, they are jealous and suspicious of their wives, but, unlike the alcoholic, the cocaine user is commonly reticent; he is not willing to talk of his troubles.

The prognosis is always bad, even in the best cases. This drug can be withdrawn from a patient more rapidly than is possible in chronic poisoning from morphine, but a relapse is to be expected.

In dipsomania, morphinomania, and other drug habits, and in the cases of vicious and degenerate children, many encouragingly good results have been reported from the use of hypnotism. Forel, Voisin, Ladame, Tatzel, Hirt, Nielson, de Jong, Liebeault, Bernheim, van Eeden, van Renterghem, Hamilton Osgood, Wetterstrand, Schrenck-Notzing, Kraft-Ebbing, Francis Cruise, Lloyd Tuckey, Kingsbury, Woods, and others have undoubtedly cured dipsomania by hypnosis.

Wetterstrand alone cured 37 of 51 cases of morphinism by hypnosis. One of these patients had been using morphine for fourteen years and morphine with cocaine for an additional four years. All his cases except one were treated at home—they were not obliged to go to a hospital or sanitarium.

As to vicious children: Liebeault in 1887 recorded 77 cases, 45 of whom were boys and 32 girls. By hypnosis 56 of these were cured, 9 improved, 12 were not affected.

As to the so-called dangers of hypnotism in the hands of skilled physicians, there are none. Forel said: "Liebeault, Bernheim, Wetterstrand, van Eeden, de Jong, I myself, and the other followers of the Nancy school, declare absolutely that, although we have seen many thousands of hypnotised persons, we have never observed a single case of mental or physical harm caused by hypnosis." Travelling mountebanks that hypnotise in public can do harm, and they should be prevented from so doing. On the continent of Europe only physicians are permitted to use hypnosis.

For a bibliography of hypnosis as a curative agent, see Allbutt's System of Medicine, vol. viii. p. 428 (The Macmillan Co.).

In Génicot's Theologiae Moralis Institutiones, vol. i. p. 162 (Louvain, 1902), is the following passage: "Videtur licitum ebrietatem inducere ad morbum depellendum, si quando practicum est, ex gr. ad typhum depellendum, vel ad coercendam vim veneni quod e serpentis morsu haustum sit (Sabetti. N. 149). Similiter, per se licebit sensus sopire ope ebrietatis ad magnos dolores levandos: nullum enim discrimen morale videtur inter hoc medium et alia, ex gr. chloroformium, quae adhiberi solent."

That is, Father Génicot permitted alcoholic intoxication to cure typhus or typhoid (typhoid is called typhus abdominalis in Europe) and snake bite, or to quiet great pain, as chloroform is used, in his opinion. This doctrine would be correct morally if from a medical point of view alcoholic intoxication cured typhus, typhoid, or snake bite, but it does not. Alcoholic liquors are necessary in some stages or forms of typhus and typhoid, and they must be administered skilfully; but to induce alcoholic intoxication in any pathological condition is always to add a grave poison to the disease already at work. The very name of the condition is intoxication, poisoning. You can end a toothache by removing a man's jaw, but the practice is not to be encouraged.

In America, when a person is bitten by a rattlesnake or copperhead, the first aid to the injured is commonly a pint of whiskey. You might better rub milk on the patient's bootheels, because the milk is harmless, but the pint of whiskey is anything but harmless; and one is as good as the other as far as curing the snake bite is concerned. Whiskey is popularly supposed to be a good medicine in all the ills of humanity. It is a good medicine in certain cases and a very bad medicine in others. A snake bite is a startling evil, and while far from a physician the early settlers gave the patient the only medicine they had, whiskey, and if a little is good a great deal is better. As the "bite" of the North American snakes is frequently not fatal, some early victims grew well in spite of the snake venom and the added whiskey poisoning; therefore a pint of whiskey cured them, post hoc ergo propter hoc. Thus the "cure" became fixed in the popular ignorance, and some moral theologians, without investigating the matter, fixed it deeper. The venom of the East Indian cobra and of other tropical and subtropical snakes would not be affected in the slightest degree by all the whiskey in Kentucky. The only hope in such cases, is in Calmette's antitoxin, administered within an hour or two after the poisoning.

Snake venom paralyses the muscles of respiration, and the patient ceases to breathe. A little whiskey may do good—whiskey pushed to intoxication is very injurious. Artificial respiration, if needed, as in a case of attempted resuscitation after partial drowning, with skilful stimulation by a physician, and the use of an antitoxin, are the main parts of the treatment in snake poisoning; but to pour a pint of whiskey into the victim is cruel ignorance. Patients often come into dispensaries showing bitten wounds which are stuffed with hair from the dog that did the biting; whiskey causes a man to see snakes, therefore use "hair from the dog that bit you." This may be good homoeopathy, but it is not medicine.

The making a man drunk with alcohol "to remove great pain" is a treatment not used by reputable physicians: there are many correct medical methods of removing pain, but a big draught of whiskey is not one of them. Even in a case where a physician can not be found, it is usually questionable whether the effect of alcoholic intoxication would not be worse than the irritation of the pain; and if it were not, where is the line to be drawn? Some male and female old ladies can work up "great pain" from a colic. The bigger and stronger a man is, especially if he has never been ill before, the greater his "agony" when he is having a tooth filled.

AUSTIN ÓMALLEY.

HEREDITY PHYSICAL DISEASE

Heredity is a very vexed question, with regard to which most varied opinions are held even by those apparently justified in having opinions, so that it is evident we are as yet only crossing the threshold of definite knowledge and are not near anything like the clear view that many people have imagined. The most striking proof of this inchoateness of scientific knowledge of heredity is the fact that within five years the work of a monk in Austria, done about forty years ago, which has lain utterly unrecognised ever since, has come to be accepted as the most striking bit of progress made—almost the only real scientific knowledge with regard to heredity that was acquired during the whole nineteenth century. Father Gregor Mendel's work [Footnote 2] was done with regard to the pea plants in his monastery garden, and it revolutionised all the supposedly scientific thinking with regard to heredity that has been current in biology for half a century.

[Footnote 2: See American Ecclesiastical Review, Jan. 1904; Walsh, A New Outlook in Heredity.]

This serves very well to show how far in advance of observed facts theories of heredity have gone. There is undoubtedly a very significant influence exerted over life and its functions by the special powers that are transmitted by heredity. How far this influence extends, however, and how much it may be said to rule details of existence, of action and in human beings, that complex of elements we call character, is entirely a matter of conjecture, and the belief in its extent, or limitation, depends absolutely on the tendency of the individual mind to accept or discredit certain theories in heredity which have had great vogue.

Until within a very few years it was considered a matter of common experience and observation that under some circumstances, at least, acquired characteristics were transmitted by heredity. That is to say, it has been definitely asserted as probable, and by many even intelligent people considered absolutely certain, that modifications of a living being undergone during the course of its existence might influence the progeny of that being in various but very definite ways. It was not, of course, thought that if a man lost an arm and subsequently begot a child, the child would be born without an arm, but slighter modifications of the organism were somehow supposed to be transmissible; and, on the other hand, modifications which affect important organic structures of the body were somehow thought to have a definite effect, by transmission, upon corresponding portions of the progeny.

When this theory is stated thus baldly, very few people confess their belief in it, yet how many there are who find ample justification for such expressions as, "His father suffered from rheumatism and it is not surprising then that he should have it"; "Her mother had heart trouble and we've always been afraid she would suffer in the same way." We are only just beginning to get beyond the period in which consumption was thought to be directly and almost inevitably inherited. With regard to mental ailments this was frankly conceded by nearly every one. If the direct ancestry suffered from mental disease of some kind, then it is not considered surprising that the immediate descendants should be mentally affected in some way. Physicians are quite as prone as those without medical training to make loose statements of this kind.

Of course there is a reason for the confusion that exists in this matter. Oliver Wendell Holmes once said that he could cure any patient that came to him for treatment, if he but applied to him in time. For proper success, however, he considered that many of his patients would have had to come to him in the persons of their great grandfathers. As a matter of fact, many of the supposed hereditary influences that are traced only to a father or a mother are family conditions that have existed many generations, and that were probably originally acquired, but the moment of whose acquisition cannot be definitely determined. We know that the Hapsburg lip has been a distinguishing feature, a persistently recurring peculiarity, in some of the members of the Austrian ruling family in nearly every generation for seven centuries. How much farther back than that it goes we have no way of determining. It is a family affair, a characteristic which became a matter of heredity perhaps ten centuries ago, but the mode of its original acquisition is a mystery.

There is no really great scientist in biology at the present moment who teaches the hereditary transmission of acquired characteristics. Modifications of the organism that become matter for heredity have existed for many generations and we cannot tell just how they began. There is no doubt that there is some hereditary influence, for insanity in the same family is likely to keep recurring in successive generations. More than this, affections of certain less important organs are evidently a common trait in certain family strains. There is no doubt that in some families stomach affections are the rule in successive generations. It is very hard to say, however, just when such defective organisation became a family trait. The tendency to nervous affections is undoubtedly a similar family affair. Certain affections have been hereditary traits for many generations. An excellent example of this is the so-called Huntingdon's chorea, which several generations of American doctors, of the name of Huntingdon, by following carefully the history of certain families on Long Island, succeeded in tracing through four generations.

The habits of life of a father or a grandfather may so weaken the physical constitution of his descendants as to make them less capable of resisting infections in the physical order, or in the moral order of withstanding trials and temptations, and the allurement to abuse of nervous excitement to which they may be subjected. That some acquired pathological condition, however, as stomach trouble, or heart trouble, or affection of the liver or of the brain, should be directly transmitted, is quite as nonsensical as that the loss of an arm should be a subject for hereditary transmission. On investigation it will be found that the pathological conditions of immediate ancestors are themselves only a manifestation of family traits that have existed for many generations. The possibility of inheritance must therefore always be borne in mind. We are utterly unable as yet to understand how such family traits are originally developed, since, in ordinary experience, at least, acquired characteristics are not the subject of inheritance or transmission, and consequently it becomes difficult to understand how they ever became impressed upon the family constitution.

Notwithstanding this general principle with regard to heredity, there are a number of striking observations which show that even unimportant peculiarities may occur from generation to generation, though it is not always easy to decide where the peculiarity originated. The well-known example of the occurrence of six toes has already been mentioned, and is an oft-quoted bit of evidence as regards hereditary transmission. An extra finger on the hand, or some portion of an extra finger, at least, comes in the same category. Not long since it was pointed out that harelip is another of these peculiarities that readily lends itself to hereditary transmission. Recently there was the report of a family into which there were born four girls with harelip and cleft palate, and three boys not showing any trace of these deformities.

Often when in such cases there is no definite history of harelip, it is found that in either one of the parents there is a very high arched palate and a thin upper lip, showing that the normal occlusion of the cleft which exists here during foetal development is not quite perfect, and this peculiarity may be traced for several generations back, with an occasional occurrence of harelip as an exaggerated example of the faulty tendency not to produce sufficient tissue in this neighbourhood for the proper closure of the embryonic cleft.

An even more striking manifestation of a physical anomaly, as a family trait, is the condition known as hemophilia. This tendency to bleed easily, so that a slight scratch, or the pulling of a tooth, may give rise to fatal hemorrhage, occurs, as a rule, only in males, but is transmitted through the female line. It is in the mother's male relatives that the history of its previous occurrence is found, and the tendency usually can be traced through several generations, until it is lost in vague tradition. It is no wonder, with such examples before them as six-toedness, harelip, and hemophilia, that physicians have been ready to accept heredity of qualities in the moral order, traits of character and disposition, and pathological tendencies to crime or passion or indulgence.

One of the most frequently discussed conditions of supposed pathological inheritance of this order is dipsomania. Everyone has heard it said, "Poor fellow, how can he help it; his father was a drunkard before him." As we have already said, in such direct cases inheritance is absolutely unproven. An alcoholic father may transmit a very weak physical constitution to his children, and this may prove inadequate to enable them to withstand the emotional strain and worry of modern great city life, and, as a consequence, they may take to alcohol for consolation until the habit is formed, and then the craving for stimulants supplies the place of any hereditary influence that may be supposed to be needed.

Of course there are cases of the drink habit in which, after a number of generations of family history of alcoholism, an individual seems to have the craving for stimulants born in him. In such cases it is not unusual to find that the patient, for such he must be considered, is able to avoid indulgence in liquor entirely, except at certain times. Every physician of any large experience has had under his care dipsomaniacs who had no difficulty in keeping away from liquor for weeks, or even months, but who had regularly recurring periods, sometimes as far apart as every three months, when they had an irresistible craving for stimulants come over them. The regularity of the interval in these cases is often very remarkable. Here, of course, we may be in the presence of some as yet not well-understood periodical law of cell life, with consequent depression, and then the irresistible craving for stimulation. As a rule, however, it would seem that in most of these cases suggestion has great influence. As we have said elsewhere, with regard to suicide, when a man has constantly before his mind's eye the fact that a father, perhaps a grandfather, or other members of the family, have committed suicide, he is likely to be much more easily led to the thought of this way of escaping hard conditions in life than are other individuals. The man who knows that the fact that his father indulged too freely in stimulants will be looked upon by many as an excuse for his deviations in this matter is likely to be more easily led to take an occasional drink at moments of depression, or for friendship's sake, though he realises that it so weakens his will power over himself that he is likely to take too much before he stops.

The passage in Julius Caesar (Act. I. sc. 2) in which Cassius says:

"The fault, dear Brutus, is not in our stars, but in ourselves,"

illustrates one phase of the subject. There are, of course, many other things besides the drink habit, with regard to which men are prone to find excuses in heredity, and to consider that somehow their ancestral tendencies make them not quite responsible for actions commonly considered the result of malice or passion, rather than hereditary influence, and our great English poet, knowing men so well, has stated the truth forcibly.

In King Lear there is an often quoted passage which properly stigmatises the opinion in this matter held by those who would find excuses for wrong-doing in hereditary qualities:

"This is the excellent foppery of the world, that, when we are sick in fortune,—often the surfeit of our own behaviour,—we make guilty of our disasters the sun, the moon, and the stars; as if we were villains by necessity; fools by heavenly compulsion; knaves, thieves, and treachers by spherical predominance; drunkards, liars, and adulterers by an enforced obedience of planetary influence; and all that we are evil in, by a divine thrusting on; an admirable evasion of whore-master man, to lay his goatish disposition to the charge of a star!"

One phase of the question of hereditary tendencies to inebriety is extremely interesting from a physiological and sociological point of view. As the result of carefully gathered statistics, there seems to be no doubt now that when children are conceived while the parents, or either of them, is in a state of drunkenness, the offspring is very likely to be of low-grade physical constitution and often of very neurotic tendencies. In France, particularly in the case of a number of insane children and idiots, histories of this nature have been obtained in confirmation of this unfortunate factor as an element in degeneracy. In general it may be said that about one-third of the admissions to homes for children of low intelligence, as well as to insane asylums, are due to this cause.

There is in this, of course, an added motive for temperance, and it would seem that parents should be warned of the danger to which they are subjecting their offspring by excessive indulgence in alcohol, when it may be followed by such serious and lasting results to the beings on whom their love and affection will be expended in the future. This phase of alcoholic excess has never been taught as insistently as its importance would demand, perhaps because of the delicacy of the subjects which it involves; but it is too significant a factor in making or marring progress in the development of the race to allow any pusillanimous motives to prevent the spread of precious knowledge. [Footnote 3]

[Footnote 3: The present conditions that obtain with regard to the celebration of marriages are very prone to have a certain amount of intoxication as their result. Perhaps, then, it is a fortunate thing, as has often been said, that the first child is not born until some considerable time after it might normally be expected. It has been said more than once, however, that first children are a little more likely to have certain degenerative defects than are others, and a connection has been found between certain abuses of stimulants and incidental exhaustion to account for this. One of the most amusing things to Li Hung Chang, on his travels through our country, was the curious publicity we give to everything connected with marriage, while presumably our Christian ideas should rather counsel a veiling of the mysteries, religious and physical, connected with it. Certain it is that the present tendency towards farewell dinners at clubs, and other festivities of various kinds, are not at all likely to result in benefit to the presumably hoped-for offspring.]

The only real light that has been thrown on the puzzling details of heredity has come from work in the same field in which Mendel made his ground-breaking observations. De Vries, the professor of botany at the University of Amsterdam, has succeeded in showing that new species of plants may be made to arise by careful attention to certain anomalous plants which occur from generation to generation. These plants breed true, that is, maintain their own peculiarities. To begin with, they are quite different from the parent plants, and the difference is perpetuated by inbreeding.

So far the problem of the origin of species has been supposed to depend upon the normal variation that is noticed in plants and animals. All living things differ from one another, even though they may belong to the same species, and differ sometimes in remarkable degrees. This continuous variation was supposed to account for the origin of new species when it became excessive. It has become well recognised now, however, that such differences gradually disappear in the course of the normal multiplication of plants and animals. The tendency is much more towards the disappearance than the maintenance of peculiarities.

There are certain discontinuous variations, however—sports, as they are called—in plants which differ very markedly in some quality from others, and these have the tendency to perpetuate themselves. Just why these sports occur is not known, nor how. They occur in a certain small percentage of all normal plants, but may die out, though it takes but little encouragement to succeed in helping them to maintain themselves. It is this that De Vries has done, and thus has succeeded in raising what would be called new species of plants.

This same thing would seem to occur in human beings. Some definite variation occurs as a consequence of a peculiar embryologic process. This becomes stamped upon the genital material and appears in the subsequent generations. It does not occur as a consequence of pathological changes nor of mere embryonic faults; it is almost as if it were something introduced from without. Once having found an entrance, however, it affects the germinal material and thus perpetuates itself.

With regard to plants, it has been suggested that the only explanation available for the occurrence of sports is that there is a purposeful introduction of them as the result of the laws of nature, and that it is thus that evolution is intentionally brought about. This is, of course, a scientific reversion to teleology once more, but the question of teleological influences has been discussed more seriously in the last few years in biological circles than ever before. Unfortunately for the coincident evolution argument involved in human beings, the peculiarities introduced, which become the subject of inheritance, do not make for the development, but rather for the degeneracy, of the race. Even such peculiarities as six toes can scarcely be said to add any special feature of advantage to man in his struggle against his environment.

It is agreed by many of our best authorities in biology, zoology, and botany, by such men as Professor Wilson of Columbia University, Professor Thomas Hunt Morgan of Bryn Mawr, Professor Castle of Harvard University, Professor Bailly of Cornell University, Professor Michael Guyer of the University of Cincinnati, Professor Spillmann, who is the Agrostologtst of the United States Government, and Professor Bateson of the University of Cambridge, England, that these principles of heredity enunciated by Father Mendel will undoubtedly revolutionise the modern knowledge of the subject. In the meantime, however, all the old theories are in abeyance. Darwin's work and Weissmann's brilliant theories and observations must give way, while the application of these new laws is being worked out to their fullest extent. While the influence of heredity can not be denied, there is undoubtedly a tendency to overestimate the influence on the physical being of the power of hereditary transmission, and, on the other hand, to underestimate the influence of this same force as regards disposition and character. There is no doubt now that the physical basis influences the exercise of the will, and that consequently responsibility is not infrequently modified by the hampering influence of unfortunate physical qualities. This truth makes for that larger charity in the judgment of the actions of others which enables physicians to realise how much men are to be pitied, while its failure of recognition by the "unco guid" not only causes suffering, but in the end adds to the amount of evil.

JAMES J. WALSH.

HYPNOTISM SUGGESTION AND CRIME

In recent years a quasi-unconscious state, induced by suggestion and called the hypnotic trance, has come to occupy a very important place in the popular mind. Hypnotism, as the general consideration of this state is known, has attracted not a little attention, as well from physicians as from those interested in psychology. The hypnotic (Greek, , sleep) trance is a condition in which voluntary brain activity is almost completely in abeyance, though the mind is able passively to receive many impressions from the external world. There are very curious limitations in the effect of the hypnotic state upon the various senses. While visual sensations, and, as a rule also, impressions from the tactile sense, lose their significance, or are translated according to the will of the person active in producing the hypnotic state, or of some person present making suggestions, auditory sensations are quite normally perceived. The patient has all the appearance of being asleep, though motions, and even locomotion, are often possible, and are performed as if the patient were walking in sleep.

The hypnotic state is a partial sleep, then, of the motor side of the nervous system and of portions of the sensory nervous system. Certain of the higher intellectual powers, however, are entirely awake, and capable of being impressed through the hearing, and thus hypnotic suggestion has a place. For a time, under the influence of Charcot and his disciples, there was a very generally accepted opinion that the hypnotic trance was a pathological condition, somewhat allied to the cataleptic phase of major hysteria. It is well known that persons suffering from severe attacks of hysteria may, while apparently unconscious, yet receive suggestions through the hearing. On the other hand, the production of cataleptic and other strained attitudes, in the maintenance of which fatigue seems to play no part, is possible by means of hypnotic suggestion in susceptible individuals.

Further investigation, however, seems to have shown that the hypnotic state is rather to be considered as a quasi-physiological condition, somewhat related to sleep, all the mystery of which is not as yet understood. This is not surprising when we realise that such a normal and absolutely physiological condition as healthy sleep is yet without a satisfactory explanation on the part of physiologists. Hypnotism is recognised now as having a certain limited power for good, though the benefit derived from it is apt to be temporary, and the operator loses his power after a time,—not so much failing to produce the hypnotic condition, as failing to have his suggestions favourably accepted by the subject While the Nancy school of hypnotism insisted that most people were susceptible to the hypnotic trance, it is now generally considered that something less than 40 per centum of ordinary individuals can be brought under its influence.

Much has been said of the dangers of hypnotism. There seems no doubt that very nervous persons are likely to be hurt by repeated recourse to the hypnotic condition. After a time they are likely to live most of their lives in a half-dreamy condition, in which initiative and spontaneous activity becomes more difficult than before. Where persons have been hypnotised by means of the flash of a bright object, or by some other special means, it sometimes happens that accidentally some similar object may send them into hypnotic trance. After a time, too, auto-hypnotism becomes possible, and much of the individual's waking time is occupied with efforts to keep himself from going into the hypnotic trance. These are, however, very extreme cases, likely to occur only in those who are not of strong mentality in the beginning. Unfortunately these are the individuals who are most likely to be made the subjects of repeated and prolonged hypnotic experimentation on the part of unscrupulous charlatans.

For the great majority of those that are susceptible to the hypnotic condition, there is very little danger. We now have on record the experiences of men who have seriously devoted many years to the study of hypnotic phenomena. There is entire agreement among these men that the possible dangers of hypnotism have been exaggerated. Indeed, it may be as well to say at once that most of what has been written with regard to the dangers of hypnotism has come from those who have least practical experience with the condition. Dr. Milne Bramwell, who, for a quarter of a century, has had a very extensive experience with hypnotism in its many phases, in his recent book on hypnotism, deliberately speaks of the "so-called dangers" of hypnotism. He has never seen any evil effects, though he has been practising hypnotism very freely on all kinds of patients for over twenty years.

It is on the experience of such serious, disinterested observers that we must rely for our ultimate conclusions as to hypnotism, rather than on the claims of pseudo-experts who like to magnify their own powers, or on popular magazine articles, or still less the Sunday newspapers, the writers for which are mainly interested in producing a sensation. It seems probable that in the next few years hypnotism will occupy a less prominent place in popular interest than it has in the recent past. Interest in hypnotism runs in cycles, reaching a maximum about once a generation, and we are on the downward swing of the last wave of popular attention to this subject.

A subject that has attracted much attention, whenever hypnotism has been under discussion, has been the possibility of crime being committed under the influence of hypnotic suggestion. The best authorities in hypnotism seem to be agreed that subjects can not be brought by hypnotic influence to perform actions that are directly contrary to their own feeling of right and wrong. The supposed exceptions to this rule are rather newspaper sensations than real compelled crimes. There is no doubt, however, that a tendency to the performance of certain wrong actions, so that the normal disinclination to their performance becomes much less than before, may be cultivated by a series of hypnotic as well as by waking suggestions. Where the individual influenced is already characterised by weakness of will in certain directions, the added weight of the motives furnished by hypnotic suggestion may prove sufficient to turn the scale of responsibility. It is probably because of such influence that a recent case in France has attracted world-wide attention.

In general, however, it may be said that normal individuals can not be brought to the commission of crime by hypnotic suggestion, and the plea of irresponsibility, for this reason, is not worthy of consideration. There are phases of this important problem, however, which require further careful study. Undoubtedly some of the so-called inherited tendencies to the commission of crime are really instances of the influence of auto-suggestion that has kept the possibility of some criminal act constantly before the mind. Some of the cases of hereditary dipsomania are almost surely of this character. Persons whose parents have been the subject of inebriety lose something of their own will power to keep away from intoxicating drink by the reflection that it is hopeless for them to struggle against an inherited tendency.

A series of cases have been reported in which suicide has occurred in successive generations in the same family at about the same time of life. There seems no doubt that suggestion must have great influence in such cases. In one well-authenticated report, mentioned in the chapter on suicides, the members of the family were officers in the German army, and the eldest son, the family representative, committed suicide within the same five years of life, in four successive generations. The last member of the family had refused to marry, because of this doom hanging over the house, and had often referred to the possibility of suicide in his own case. In his early years he seemed to have the idea that he might escape the family fate, but after middle life he settled down irretrievably to the persuasion that he would inevitably go like the others.

Here, in America, a rather striking example of this has recently been the subject of sensational newspaper reports. A notorious gambler, whose career had seen many ups and downs, finally found himself in a condition where, strange as it may seem, legal restriction made it impossible for him to continue his usually lucrative profession. Three members of his immediate family, two brothers and his mother, had committed suicide. To friends he had sometimes spoken of this sad history of family self-murder, but always with a calm rationality which seemed to indicate that he hoped to avoid any such fate. When well on in years, however, with his means of livelihood taken from him, he, too, took the family path out of difficulties and shot himself at the door of the man who had been most instrumental in taking away from him his occupation. It seems not unlikely, from the circumstances of the case, that a double crime, homicide, as well as suicide would have been reported, only for the fortuitous circumstance that the other man was not in at a time when usually he was to be found at his office.

In such cases as these it seems reasonably clear that long-continued familiarity with a given idea produces an auto-suggestion which finally overcomes the natural abhorrence even of suicide. Something can be done for such unfortunates by suggestion in the opposite direction, and by taking care that as far as possible they are not allowed to brood over the fate they consider impending. At times of stress and emotional strain, relatives and friends must be particularly careful in their watch over them. It is never advisable that they should take up such professions as those of broker or politician, or speculator, since the emotional states connected with such occupations are likely to prove too much for their mental equilibrium.

Practically all physicians that have given any attention to the subject are convinced that not a few of the suicides, which are now so alarmingly on the increase in this country, are due to the frequent reading in newspapers of the accounts of suicides. As we have said elsewhere, brooding over the details of these is very likely to lessen the natural abhorrence of self-murder in persons that are predisposed, by melancholic dispositions, to such an act. The publication of cases of suicide can do no possible good, while it undoubtedly does, in this way, work incalculable harm. This is especially true with regard to suicides among young people, that is, individuals under twenty-five years of age. The saddest feature of recent statistics with regard to suicide is that this crime has become proportionately much more frequent among young men and young girls, and even children, than it was two or three decades ago. It has been noted, too, in many cases that a previous suicide in the family seems to have familiarised the young mind with the idea of self-destruction and thus suggested its commission.

On the other hand, among young people especially, it has been noted that there is frequently an imitative element in suicides. Three or four suicides, practically with the same details, will occur, within a few days of each other. Suicides at all ages are especially likely to occur in groups, and are often cited to exemplify the truth of the old axiom that evils never come singly. It is especially among young people, however, that this relationship to previous suicides can be traced, and there is no doubt that it is the unfortunate publicity given to suicide, with the consequent suggestive influence, which constitutes the most important factor in these cases. All the influence that clergymen can exert, then, must be wielded to suppress this, as well as the many other evils which flow from sensational journalism.

JAMES J. WALSH.

UNEXPECTED DEATH

Unexpected death and its problems constitute the principal reason why there should be a pastoral medicine, and why the clergyman must keep himself in close touch with advances in medicine. To have an ailing member of a congregation die unexpectedly, that is, without the rites of the Church, when perhaps there has been some warning as to the possibility of such an accident, can not but be a source of the gravest concern in pastoral work. Sudden death can be anticipated in many diseases that are acute, while in chronic forms of disease the sufferer can be prepared for its possibility by the administration of the sacraments at regular intervals. There is, however, an old proverb which says that death always comes unexpectedly; and even with all the modern advance in medicine, this still contains a modicum of truth. As an unprepared death is an occasion of the most poignant regret to the friends of the deceased and to the attending clergyman, it is with the idea of furnishing some data by which the occurrence of death without due anticipation may be rendered more infrequent, that the following medical points on the possibilities of a fatal termination in certain diseases have been brought together. Unfortunately, even with all our progress in modern medicine, they must be far from adequate for all cases.

Needless to say, the only rational standpoint in this matter must be that it is better to be sure than to be sorry. The impression is very prevalent now that at least the sacraments of Penance and the Holy Eucharist should be administered to the sick whenever there is even the possibility of a fatal termination of the illness. Extreme Unction is more usually delayed until there is some positive sign of approaching dissolution. Delay in its administration, however, not infrequently leads to this sacrament being given when the patient is unable to appreciate its significance. This would seem to be very far from the intention of the Church. The idea has been constantly kept in mind, then, so to advise the clergyman with regard to the liability of a fatal termination as to secure, if possible, the administration of Extreme Unction while the patient is still in the full possession of his senses.

Assured prognosis, that is, positive foresight as to the course of any disease, is the most difficult problem in medicine. Nearly 2400 years ago, when Hippocrates wrote his chapter on the progress of diseases, he stated that the hardest question to answer in the practice of medicine is, will the patient live? That special chapter of his book remains, according to our best authorities, down even to our own day, a valuable document in medical literature. It can be read by young or old in medical practice with profit. While our knowledge of the course of disease has advanced very much, the wise old Greek physician anticipated most of the principles on which our present knowledge of prognosis is founded. This fact in itself will serve to show how unsatisfactory must be any absolute conclusion as to the termination of any given disease. Our forecasts are founded on empirical data,—that is, they are the result of a series of observations,—and the underlying basis of all the phenomena is the individual human being, whose constitution it is impossible to know adequately, and whose reaction to disease it is impossible, therefore, to state with absolute certainty.

With this warning as to the element of doubt that exists in all prognosis, we may proceed to the consideration of certain organic affections which make sudden death frequent.

At the beginning of the present century, Bichat, a distinguished French physician who revolutionised medical practice, said that health and the favourable or unfavourable termination of disease depends on the condition of three sets of organs—the brain, the heart, and the lungs. This was what he called the vital tripod. It was not until nearly thirty years after Bichat's death that Bright, an English physician, taught the medical profession to recognise kidney disease. Since his time we have learned that even more important than Bichat's vital tripod, as regards health and the termination of disease, is the condition of the kidneys. We shall consider affections of these four organs, and their influence on the human system and intercurrent disease, in the order of their importance.

When kidney disease exists the individual's resistive vitality is much lowered. The kidneys are the organs which serve to excrete poisons that find their way into the circulation. When the kidneys fail to act, these poisons are retained. As a result other important organs, notably the nervous system and the heart, suffer severely because of the irritating effect of the retained poison. A patient with kidney disease runs a very serious risk in any infectious fever, no matter how mild, and such patients should always be completely prepared for a fatal termination when they acquire any of these diseases.

Nephritic patients bear operations very badly. The shock to the nervous system incident upon operation always throws a certain amount more than usual of excrementitious material into the circulation. Diseased kidneys do not fulfil their function of removing this at once, and the result is an irritated and fatigued nervous system. Anaesthetics, that is, chloroform and ether, are not well tolerated when nephritis exists, and this adds to the danger of operation in such patients. No matter how simple or short the operation that is to be performed on a person suffering from kidney disease, if an anaesthetic is to be administered it would be well to prepare the patient for an untoward event that may occur.

Kidney disease is often extremely insidious. It may develop absolutely without the patient's knowledge, even though he might be deemed to be in a position to have at least some suspicion of its existence. The story is told of more than one professor of medicine who has presented his own urine to his class for examination in order that they might have the opportunity of studying normal urine, only to find to his painful surprise that albumen was present and that he was the subject of latent Bright's disease. In these cases it is impossible to foresee results. They constitute a large number of the cases in which patients, seemingly in good health, succumb rather easily and unexpectedly to some simple disease, like grippe or dysentery. It is well to take the precaution, then, to ask the attending physician what the condition of the kidneys is in such cases. If there are anomalous symptoms, this precaution becomes doubly necessary. Even such simple infectious diseases as mumps or chicken-pox may cause a fatal issue where the kidneys are not in a condition to do their normal work of excretion.

An important class of cases for the clergyman are those which are picked up on the street. As a rule, these patients are comatose because of the presence of kidney disease. A certain proportion of them are unconscious because of apoplexy. Very often the patients have had some preliminary symptoms of their approaching collapse, though these were not sufficient to make them think that any serious danger threatened. As a consequence, they will not infrequently have had recourse to some stimulant. It seems unfortunately to be almost a rule, when such cases are picked up, if there is the odour of alcohol on their breath, to consider that the condition is due to alcoholism. Every year, in our large cities, some deaths are reported in the cells of the station houses because a serious illness was mistaken for alcoholism as a result of the odour of the breath. Needless to say, then, the odour of alcohol on the breath of a person in coma should not deter a clergyman from waiting for a time to be sure his ministrations may not be needed for something much more serious than alcoholism.

Patients suffering from kidney disease bear extremes of cold and heat very badly. In cold weather the fact that the blood is driven from the surface of the body lessens the excretory function of the skin, and this throws the work of this important organ, so helpful an auxiliary in excretion, back upon the kidneys. Besides, congestions of internal organs are not infrequent during cold, damp seasons, and these bring on exacerbations of previously existing ailments that may make fatal complications. In summer intense heat leads to many more changes in the tissues, and so provides more material to be excreted than in temperate weather. Patients picked up on the street, then, at such time, will usually be found to be suffering from kidney disease. Though in profound coma, such patients seldom die without recovering consciousness. Not infrequently, after the primary stroke of the coma, there is, in an hour or two, a period in which the patient becomes almost completely rational. This period of consciousness does not last long, in many cases, and should be taken immediate advantage of, yet without unduly disturbing the patient.

There is a well-known tendency in kidney disease to the production of oedema, that is, to the outflow of the watery constituents of the blood into certain loose tissues of the body. This is easily recognised, and constitutes a valuable sign of kidney disease in the swelling of the eyelids and of the feet, that occurs so often in patients suffering from kidney trouble. The usual rule is, if the oedema begins in the face, it is due to the kidneys; if in the feet, to the heart. The cause in the latter case is the sluggish circulation due to the weakness of the heart muscle, which delays the blood so long in the extremities that its watery elements find their way out into the tissues. In kidney disease this tendency to oedema constitutes a distinct danger that may involve sudden death in certain affections. In patients suffering from kidney disease any acute sore throat involving the larynx and causing hoarseness may be followed by what is called oedema of the glottis. This is often fatal in a very short time. The glottis is the opening between the vocal cords through which respiration is carried on. This opening is but small, and swelling of the surrounding tissues readily encroaches upon it, and soon causes difficulty of breathing. If the swelling is not relieved without delay, death takes place from asphyxiation. This was probably the cause of death in George Washington. In almost the same way any acute affection of the lungs that occurs in a patient suffering from kidney disease may be followed by oedema of the lungs. The outflow of serum from the blood vessels into the loose tissues of the lungs so encroaches upon the space available for breathing, and at the same time so reduces the elasticity of lung tissue, that respiration becomes impossible, and death takes place in a few hours. This is often the cause of unexpected death after operations. The kidney affection in the patient is so slight as to have been unsuspected, or to have been considered of not sufficient importance to render the operation especially dangerous.

After kidney disease the most important factor in the production of unexpected death is heart disease. In about 60 per centum of the patients who die suddenly, in the midst of seemingly good-health, death is due to heart disease. All forms of heart disease may be considered under two heads—the congenital and the acquired. The congenital form of heart disease usually causes death in early years. If such patients survive the fourth or fifth year, they are usually carried off by some slight intercurrent disease shortly after puberty. A few cases of congenital heart disease, however, live on to a good old age and seem not to be seriously inconvenienced by their heart trouble. Most of the acquired heart disease, that is, at least 65 per centum of it, is due to rheumatism. All of the infectious fevers, however, may cause heart disease, and scarlet fever especially is prone to do so; heart complications occurring in about one out of every ten cases. The probabilities of sudden death in a case of heart disease depend on what valve is affected and what the condition of the heart muscle is. Most of the cases of sudden death occur in disease of the aortic valves, that is, of the valves that prevent the blood from flowing back from the heart after it has been pumped out. Diseases of the other side of the heart, the mitral valve, cause lingering illness until the heart muscle becomes diseased, when sudden death usually closes the scene.

Diseases of the aortic valves of the heart cause visible pulsations of the arteries, especially of those in the neck. This readily attracts attention if one is on the lookout for it. Deaths in heart disease, whether sudden or in the midst of apparent health, or as the terminal stage after confinement to bed because of weak heart, are apt to occur particularly during continued cold or hot spells. Each of the blizzards that we have had in recent years has been the occasion for a markedly increased mortality in all forms of heart disease. The cold itself is exhaustive, and the heavy fall of snow, by delaying cars and modes of conveyance generally, is very apt to give occasion for considerably more exertion than usual. Besides, cold closes up the peripheral capillaries and makes the pumping work of the heart much harder than before. At times of continued cold, in our large cities particularly, the ordinary arrangements for heating the house fail to keep it at a constant temperature, and this proves a source of exhaustion to cardiac patients.

Heated spells, if prolonged, always cause an increased mortality in such patients, because heat is relaxant and this leads to exhaustion. Patients who have been nursed faithfully through a severe winter will sometimes succumb to the first few successive days of hot weather that are likely to come at the end of May or the beginning of June. The deaths that occur during the hot spells of July and August are more looked for and accordingly prove not so unexpected.

The warning symptom in heart disease that the patient is giving out is the development of irregularity and rapidity of the pulse. On the other hand, when a pulse has been running rapidly for weeks and then drops to below the regular rate, to 50 or 60, a fatal termination may be looked for at almost any time, though, of course, the patient may rally. The prognosis of heart cases is extremely difficult. Confined to bed and evidently seriously ill, they may continue in reasonably good condition for months, and then some indiscretion in diet, which causes a dilation of the stomach with gas, pushes the diaphragm up against the heart, adds a mechanical impediment to the physical difficulties the organ is already labouring under, and a sudden termination may ensue. As a rule, lingering heart cases terminate suddenly and often with little warning of the approach of death.

An interesting set of heart symptoms, for the physician as well as the clergyman, are those which occur in what is called angina pectoris, heart pang, or heart anguish. Serious angina pectoris occurs in elderly people whose arteries are degenerate. Its main symptom is a feeling of discomfort which develops in the praecordia,—the region over the heart. This discomfort may often increase to positive cutting pain. The pain is often referred to the shoulder, and runs down the left arm. This set of symptoms is accompanied by an intense sense of impending death. When the patient's arteries are degenerated, this train of symptoms must always be considered of ominous significance. A readily visible sign of arterial degeneration can sometimes be noted in the tortuous prominent temporal artery just above the temple.

Heberden, an English physician, a little over a century ago, pointed out that there existed in cases of true angina pectoris a degeneration of the coronary arteries. These are the arteries which supply the heart itself with blood. As might naturally be expected, their degeneration seriously impairs the function of the heart muscle. The first patient in whom the condition was diagnosed during life was the distinguished anatomist, John Hunter. Hunter was of a rather irascible temperament, and after he had had several of these attacks, and a consultation with Heberden convinced him of their significance, he is said to have remarked, "I am at the mercy of any villain who rouses my temper." Sure enough, Hunter died in a sudden fit of anger within the year after making the remark. Charcot, the distinguished neurologist, suffered from attacks of angina pectoris, and was asked by his family to consult a distinguished heart specialist for them. He said: "Either I have degenerated heart arteries, or I have not. I believe that I have not, and that my attacks are due to a nervous condition of my heart. If I should consult the physician you mention, and he were to tell me that my attacks are due to degeneration of the heart, he would advise my giving up work. That I am not ready to do, and so I prefer to take my own assurance in the matter." A few years later he was found one morning dead in bed. In many of the cases of death in bed, especially where some complaint of pain has been heard during the night, death is due to that condition of the heart arteries which causes angina pectoris, though it may be the first attack which proves fatal.

There is a condition similar to angina pectoris, sometimes called pseudo-angina, or false heart pang, which occurs in individuals from fifteen to thirty years of age. It is often a source of great worry. It occurs in young persons of a nervous temperament who have been overworked or overworried and have run down in weight. There are always accompanying signs of gastric disturbance. The casual factor of the symptoms seems to be a more or less sudden dilation of the stomach with gas. As the stomach lies just below the heart, only separated from it by the comparatively thin layer of diaphragm, the heart is pushed up and its action interfered with. In healthy individuals this causes no more than a passing sense of discomfort and some heart palpitation. That it is which sends so many young patients to physicians with the persuasion that they have heart disease, when they have nothing more than indigestion. In nervous individuals, however, this interference with the heart action disturbs the nervous mechanism of the heart, which is very intricate and delicate, and gives rise to the symptoms of false "heart pang." One of these symptoms is always, as in true angina pectoris, an impending sense of death. This can not be shaken off, and is not merely an imagination of the patient. Pseudo-angina is, however, not a dangerous affection. Patients can usually be assured that there is no danger of death. This assurance is not absolute, however. For some of these cases have congenital defects in their coronary arteries, and the nervous system of the heart itself, which make them liable to sudden death. It is sometimes impossible to differentiate such cases of organic heart defects from the ordinary functional heart disturbance due to indigestion, which causes simple curable pseudo-angina. Young patients may usually be disabused of their nervousness in the matter, but absolute assurance can not be given until the case has been under observation for some time.

After the heart, the head is the most important factor in sudden death. The most frequent form of death from intra-cranial causes is apoplexy. Apoplexy, as the name indicates—a breaking out—is due to a rupture of one of the arteries of the brain, and a consequent flowing out of blood into the brain tissue. The presence of the exuded blood causes pressure upon important nerve tracts, and so gives rise to unconsciousness, to paralysis, and to the other symptoms which are noted in apoplexy. There are a number of symptoms that act as warnings of the approach of apoplexy. First, it occurs only in those beyond middle life, that is, in individuals over forty-five, and in these only where there is marked degeneration of arteries. The degeneration of the arteries can be easily noted, as a rule, in other parts of the body. The condition known as arterio-sclerosis, that is, arterial hardening, can be detected by the finger at the wrist, or by the eye in the branch of the temporal artery, which can so frequently be seen to take its sinuous course on the forehead behind and above the eye. At the wrist the thickened artery is felt as a cord that can be rolled under the finger. It is not straight as in health, but is tortuous, because the overgrowth in the walls, which makes it thick, has also made it longer than normal, thus producing tortuosity.

Besides these objective signs, as they are called, there are certain subjective signs, that is, signs easily recognised by the patient himself, which should put him on his guard, and at the same time serve as a warning to the clergyman, should he hear of their presence. These signs are recurring dizziness, or vertigo, not clearly associated with gastric disturbance; tendency of the limbs, and especially the fingers and toes, to go to sleep easily, and when there is no external cause for this condition; tendency to faintness and to dizziness when the patient rises in the morning, especially if he assumes the erect position suddenly; tendency to vertigo when the patient stoops, as to tie a shoe, or pick up something from the floor, and the like; finally, certain changes in the patient's disposition, with a loss of memory for things that are recent, though the memory may be retained for the happenings of years before. When several of these symptoms occur, patients who are well on in years should take warning of the fact that they are liable at any time to have a stroke. Needless to say, this has no reference to the cases of young nervous persons who may readily imagine that they have some or all of these symptoms. Apoplexy is typically the disease of those over fifty years of age.

There may even occasionally be slight losses of power in the hand or foot that point to the occurrence of small hemorrhages in the brain, that is, slight preliminary "strokes."

Patients that have had these symptoms should not, as a rule, be allowed to leave home unattended. If the apoplexy occurs in the street they are liable to be mishandled by those ignorant of their true condition. The clergyman is usually summoned at once in these cases and may reach the stricken individual before the physician. Some words, then, with regard to the general management of such patients will not be out of place. As a rule, when a patient is taken with some sudden illness which causes him to fall down unconscious, the first thing done is to dash water in his face, force a stimulant down his throat, put his head low down, and loosen the clothing around his neck. Most of these proceedings are the very worst things that could be done for a patient suffering from apoplexy. The rough handling, particularly, and the administration of a stimulant, will surely do harm. The water on the face will certainly do no good.

Apoplectic patients can be recognised from those who are merely in a fainting fit, first, by the fact that they are usually old, while the fainters are young; and secondly, by the manner of the breathing. In a faint the breathing is shallow and faint, not easily seen. In apoplexy it is apt to be deep and long. It may be irregular, and it is always accompanied by a blowing outward and inward of the cheeks, and especially of the side of the face which is paralysed, as a consequence of a hemorrhage into the brain.

The lips are forced outward and drawn inward during the respiration. In such cases the patient should be moved as little as possible; stimulants should be avoided, and the head should be placed higher than the rest of the body, so as to make the hemorrhage into the brain as small as possible, by calling in the assistance of gravity to keep the heart from sending too much blood into the head. Besides this placing the head high, there is only one other helpful measure that even the physician can practise, except in rare cases, that is, to put an ice-bag on the head. For this a cloth dipped in cool water may be used in an emergency. Of course, as soon as the doctor arrives, the patient should be left entirely to his care.

The artery that ruptures in the brain, in cases of apoplexy, is practically always the same. Its scientific name is the lenticulo-striate artery, but it is oftener called by the name given it by Charcot—the artery of cerebral hemorrhage. The reason why arteries in the brain rupture rather than arteries in other organs is that in the brain, in order to avoid the demoralising effect of too sudden changes of blood pressure upon the nervous substance, the cerebral arteries are terminal, are not connected directly with a network of finer arteries as in the rest of the body, but gradually become smaller and smaller, and end in the capillary network which is the beginning of the venous vascular system. This special artery ruptures, because it is almost on a direct line from the heart, and so blood pressure is higher in it than in other brain arteries.

The tradition that people with short necks are a little more liable to apoplexy than are those of longer cervical development has a certain amount of truth in it, though not near so much as is often claimed for it. Another predisposing element to apoplexy is undoubtedly heredity. Families have been traced in which, for five successive generations, there have been attacks of apoplexy between fifty-five and sixty years of age. Short-necked people, with any history of apoplexy in the family, should especially be careful, if they have any of the symptoms—dizziness, sleepy fingers, etc.—that we have already noted.

There is a tradition that the third stroke of apoplexy is always fatal. This is without foundation in experience, though of course the liability of death increases with each stroke, and few patients survive the third attack. I remember seeing in Mendel's clinic, in Berlin, a man who was suffering from his seventh stroke and promised to recover to have another. Each successive stroke is much more dangerous to life than the preceding one, however. In general, the prognosis of an apoplexy, that is to say what the ultimate result will be, is impossible. The patient may come to in an hour or two, and may not come out of the coma at all. There is no way of deciding how large the artery is that is ruptured, nor how much blood has been effused into the brain, nor how much damage has been done to important nerve centres. Nor is there any effective way of stopping the effusion, though certain things seem to be of some benefit in this matter. We can only wait, assured that, in most of the cases, the patient will have a return of consciousness, at least for a time.

Next to apoplexy, injuries of the head are most important. The symptoms presented by the patient will often be nearly the same as those of apoplexy. If the skull is fractured, and the depressed bone is exerting pressure upon the brain substance, there is a similar state of affairs to that which exists in apoplexy. Any return to consciousness must be taken advantage of for the administration of the Sacraments. As a rule, it is impossible to tell the extent of the injury or to forecast the ultimate result.

A very characteristic set of symptoms develops sometimes after injuries in the temporal region or just above it. For a short time up to an hour or two after the injury, the patient is unconscious. Then he comes to for a while, but relapses into unconsciousness, from which he will usually not recover except after an operation. The explanation of this succession of symptoms is that the primary unconsciousness is due to shock—concussion or shaking up of the brain. The injury has, however, also caused a rupture of an important artery which occurs in one of the membranes of the brain in this region, the middle meningeal artery. During the state of shock blood pressure is low and hemorrhage is not severe. When consciousness is regained, blood pressure goes up and the laceration of the middle meningeal artery, already spoken of, provides an opening for the exit of considerable blood, which clots in this region and presses upon the brain, causing the subsequent unconsciousness. As a rule, the patient's only hope is in operation with ligature of the torn artery. The condition is always very serious, and complete precautions as to the possibility of fatal termination should be taken, as soon as consciousness is regained after the blow, in any case where the head injury has been severe enough to cause more than a momentary loss of self-possession. No one can tell whether there may be further change or not, and if this happens it will be in the form of an unconsciousness gradually deepening until relieved by operation or ended by death.

Tumours of the brain often produce death, but usually give abundant warning of their presence. The symptoms by which the physician diagnoses the presence of a brain tumour are vertigo, headache, vomiting, usually some eye trouble, and frequently some interference with the motion of some part of the body, because of pressure exerted upon the nerve centres which preside over its motions. Brain tumours are especially liable to develop in two classes of cases—in patients who are suffering from tuberculosis in its terminal stages or from syphilis. Where patients are known to have either of these diseases and present any two of the symptoms of brain tumour that I have mentioned, it is well to suggest at least the preliminary preparation for a fatal termination. Sometimes states of intense persistent pain, or of mental disturbance, develop in these cases and make the administration of the Sacraments unsatisfactory.

Meningitis is a fatal affection which sometimes causes sudden death, but more frequently produces unconsciousness without very much warning, and the unconsciousness lasts until the death of the patient. Meningitis is seen much more frequently in children than in the adult. Ordinarily it is due to tuberculosis. Sometimes, however, there are epidemics of cerebrospinal meningitis—spotted fever, as it used to be called. In about one-half the cases this affection is fatal. Unfortunately this disease gives very little warning of its approach in many cases before unconsciousness sets in. We have had renewed epidemics of the disease in the eastern part of the United States in recent years, and the affection is likely to occur more frequently for some time to come. The first hint of the onset of the disease during an epidemic should be the signal for the administration of all the rites of the Church.

Of late years we have learned that the pneumococcus, that is, the bacterium which causes pneumonia, may produce a fatal form of meningitis. The first symptom of meningitis is usually a stiffness of the muscles at the back of the neck. If this stiffness becomes very marked in a patient suffering from tuberculosis, or who has, or has recently had, pneumonia, or at a time when there is any reason to suspect that epidemic cerebrospinal meningitis exists in a neighbourhood, the prognosis of the case is always very serious. Every precaution should be taken to prepare the patient for the worst. Unconsciousness may ensue at any moment and no opportunity for satisfactory administration of the consolations of religion be afterwards afforded.

While Bichat put the lungs down as one of the vital tripod on which the continuance of life depends, affections of these organs very seldom lead to sudden or unexpected death. Pulmonary affections usually run a very chronic course. Acute bronchitis, however, occurring in a patient with kidney trouble, may lead to the development of oedema of the lungs, and death will usually ensue in a few hours. It may be well to note here that individuals who have what are called clubbed fingers, or as the Germans picturesquely put it, drumstick fingers, that is, fingers with bulbous ends, the finger beyond the last joint being larger than the preceding part, nearly always have some chronic affection within the thorax. This means that there is some organic affection of the heart or lungs which has lasted for many years. The existence of such condition makes them distinctly more vulnerable to any serious intercurrent disease, and this sign alone may be enough to put the attending physician on his guard as to the possibility of

fatal complications in the case.
    JAMES J. WALSH.

UNEXPECTED DEATH IN SPECIAL DISEASES

Besides the general systemic conditions in which sudden death may occur without anticipation, there are certain specific diseases of which unexpected death is sometimes a feature. For the clergyman to know the condition in which the sudden fatal termination is liable to occur is to be forearmed against the possibility of death without the Sacraments, or their enforced administration in haste, when the recipient is in a very unsatisfactory condition of mind and body. It has been said that if a normally healthy individual reaches the age of twenty-five he is reasonably sure to live to a good old age, provided he does not meet with an accident or catch typhoid fever or pneumonia.

Pneumonia is an extremely important affection as regards its prognosis. From 15 to 20 per centum of sufferers from the disease die; that is to say, about one in six of those attacked by the disease will not recover. It is a little more fatal in women than in men. It is especially serious for the very young and the old.

Healthy adults in middle life very rarely die from the disease. The prognosis of any individual case, it has been well said, depends on what the patient takes with him into the pneumonia. Serious affections of important organs nearly always cause fatal complications. If the heart is affected before the pneumonia is acquired, then the prognosis is very unfavourable, and a fatal termination is almost inevitable. If the kidneys are seriously diseased beforehand, death is almost the rule. Pneumonia developing during the course of pregnancy is fatal in more than one-half of the cases. At one time it was suggested that premature delivery of pregnant pneumonia patients might save at least the mother's life. Experience in Germany, however, has shown that, far from making the prognosis more favourable, the induction of premature labour makes the outlook a little worse for the patient. Previous affections of the lungs, emphysema, or tuberculosis, are prone to make the prognosis of pneumonia much more unfavourable than under ordinary circumstances.

Deteriorated conditions of the blood, anaemia, chlorosis—such as occurs so commonly in young women—is prone to make the outlook in pneumonia more serious. Pneumonia of the upper lobes of the lungs is more apt to be followed by complications, and is therefore more serious than pneumonia of the lower lobes. Secondary pneumonia—that is, inflammation of the lungs which develops as a complication of some other disease—is much more unfavourable than primary pneumonia which develops in the midst of health. The amount of lung involved is of course a serious factor in the prognosis. If the whole of one lung is consolidated, or if considerable portions of both lungs are thus affected, the prognosis becomes extremely unfavourable.

In persons of alcoholic habits the result of pneumonia is always to be dreaded. The more liberal has been the consumption of alcohol, as a rule, the less hope is there of a prompt, uncomplicated recovery. Stimulants are of the greatest importance in pneumonia, and the less the patient has taken of them before the development of his pulmonary affection the more effective are they when the crisis of the disease comes. The less the alcohol that has been taken habitually before the development of pneumonia, the more surely will it do the work expected of it during the course of the pneumonia. It must be borne in mind that cases of pneumonia that occur in institutions, asylums, hospitals, and the like, and in crowded quarters in tenement houses or lodging houses, have a distinctly worse prognosis than those treated in private houses, and the priest must accordingly be more on his guard and give the Sacraments early.

In pneumonia, as in typhoid fever, so-called walking cases always have a serious prognosis. They occur in very strong patients who resist, not the invasion of the disease, but its weakening influence, and keep on their feet for several days, despite the presence of symptoms that require them to be in bed. When a patient walks into a doctor's office in the third or fourth day of a pneumonia with most of one lung consolidated, exhaustion of the heart and of the nervous system, under these unfavourable conditions, will usually have made his resistive vitality very low. Such cases should be given the Sacraments early, while in the full possession of their senses. Conditions sometimes develop rather unexpectedly in which the administration of the Sacraments becomes unsatisfactory, because of the collapsed state of the patient.

This same advice holds with regard to walking cases of typhoid fever. Where strong patients suffering from the disease have insisted on being around on their feet for from six to ten days at the beginning of the affection, the prognosis becomes very unfavourable. Complications, such as hemorrhage or perforation of the intestine, occur about the beginning of the third week, and often prove fatal. All typhoid fever patients should receive at least the Sacraments necessary to give a sense of security to the priest and their friends during the course of the second week, even though they may seemingly be in excellent condition. When typhoid fever is fatal the complications occur suddenly, often without much warning; and if intestinal perforation, for instance, takes place, the peritonitis which develops makes the patient's condition very unsuitable for the reception of the Sacraments in a proper state of mind.

Typhoid fever patients sometimes die suddenly in collapse when they are convalescent. The toxine of the typhoid bacillus often affects the heart, and causes what is called cloudy swelling of its muscular fibres. This decreases very notably their functional ability. Any sudden exertion, even sitting up in bed, may cause the heart to stop under such circumstances. The modern custom in hospitals is not to allow typhoid patients to sit up in convalescence until the head of the bed has been raised gradually for several days so as to accustom the heart to pumping blood up the hill to the brain. Priests must be careful, then, when they call to see convalescent typhoid patients, not to permit them to sit up to greet them. The doctor's directions in this matter should be followed very carefully.

This sudden fatal collapse may occur after any of the infectious diseases. It is seen not infrequently after diphtheria. It occurs more rarely after scarlet fever, and even after some of the milder children's diseases. In rheumatism, especially where a heart complication has occurred, this rule with regard to sudden movements is extremely important Rheumatism is itself not a fatal disease, yet there are certain cases in which very high temperature sets in, causes delirium, and death ensues at times before the patient recovers consciousness. Where rheumatic patients show a tendency to run high temperatures, that is, 104° or higher, it is well to be prepared for this emergency.

Appendicitis is very much talked about in our day; but the fatal affection represented by the new word is no more frequent than it was half a century ago, or, for that matter, twenty-five centuries ago. People died of inflammation of the bowels and peritonitis then; and as the appendix was not known as the origin of the trouble, the fateful name was not the spectre that it is now. Practically all abdominal colic—and this means 90 per centum of all the acute pain which follows gastro-intestinal disturbance in young or middle-aged adults—is due to appendicitis. It comes on, as a rule, in the midst of good health. It is very treacherous, and when the patient is apparently but slightly ill, a sudden turn for the worse may assert itself, and an intensely painful and prostrating condition develop. Where symptoms of appendicitis are present, it is the part of safety to have the patient receive at least the Sacraments of Penance and the Holy Eucharist. When peritonitis develops, vomiting is the rule. Hence the advisability of prompt administration of Holy Communion. Extreme Unction can be given with some satisfaction, even during the disturbed period which follows a beginning peritonitis. For the peritonitis that sometimes results from appendicitis there is no hope of recovery except by operation. Operation, to be successful, must follow the perforation of the appendix not later than by a few hours.

Early pregnancy, that is, the first eight to ten weeks of gestation, is sometimes complicated by a set of symptoms the most prominent of which are sudden very acute pains in the lower part of the abdomen, followed by intense prostration, and then by the symptoms of internal bleeding,—namely, a soft pulse, pallor with cold extremities, sighing respiration, and marked tendency to faintness. When symptoms like these occur during the first three months of pregnancy, they signify, almost without exception, rupture of an extrauterine gestation-sac. Except where operation can be performed at once, these cases are almost invariably fatal. Extrauterine pregnancy occurs with greatest frequency in women who, having had one or more children, then have a period of five or more years without children, followed by pregnancy. Undoubtedly, extrauterine pregnancy, the knowledge of which is the result of medical advance in very recent years, and appendicitis, which is the growth of the last twelve years, were prominent factors in the production of many inexplicable deaths in history. These were not infrequently set down as due to poison.

Acute indigestion in elderly people is sometimes followed by sudden death. Observations in this matter have somehow become much more frequent of late years, and many of the so-called cases of heart failure belong to this group. The important nerve trunk that carries nervous fibres to the heart bears fibres to the digestive tract, the oesophagus, the stomach, the intestines, the liver as well, and also to the larynx and lungs. There is a certain intercommunication between the impulses which pass along these various nerve fibres. Intense irritation of the nerve endings in any one of these organs may be reflected back upon the heart. Curiously enough the nerve fibres to the heart that run in this trunk are many of them inhibitory; that is to say, they lessen the function of the heart or cause it to stop beating entirely. If an intense nervous irritation is set up in the stomach, reflex nervous impulses may cause the heart to stop completely and never resume its work.

Typical cases of this kind often occur during the first cold days of the winter time. Elderly people come to their meals cold and chilly, yet with appetite increased by the bracing air. They sit down at once, take a larger meal than usual, and then develop severe gastritis during the night. This is relieved by purging and vomiting, and the pain yields to the administration of morphine. Their condition improves and all danger seems past, when, on sitting up suddenly the next day, or, if left alone, getting up to get something for themselves, they collapse and are dead before help can come to them. Deaths like this sometimes occur in dysentery also, the reason being the intense nervous reflex from the irritated intestinal nerve endings which exerts its influence upon the heart nerves.

Certain diseases practically always end in sudden death and must be taken special care of by the priest for this reason. Aneurism, for instance, is one of these. An aneurism is a widening or dilatation at some point of an artery. The most important aneurisms occur in the arch of the aorta, that is, in the large curved artery which comes directly from the heart itself and of which all the other arteries are branches. Aneurisms develop, according to the expression of a distinguished American physician, in the special votaries of three heathen divinities, Vulcan, Bacchus, and Venus,—that is, in those who have worked too hard, in those who have drunk too hard, and in those who have devoted themselves too much to the pleasures of the flesh. The most important factor of all is, however, the contraction of venereal disease, especially of that form known as syphilis.

The termination of aneurism cases is usually by rupture with profuse hemorrhage. Death takes place in a moment or two. Aneurisms often cause intense pain, which is sometimes thought to be rheumatic in origin. If the aneurism, in its enlargement, meets with bony structure, it produces absorption of the bone by pressure upon it and so finds a way even through the bone to the overlying skin. This process is always intensely painful, and shortly after the aneurism appears at the surface the pressure upon the skin causes it to become thin and the aneurism may rupture externally.

Addison's Disease always ends suddenly. This is a rare affection, described by Addison, an English physician, some fifty years ago, which develops in individuals whose suprarenal capsules are degenerated. The suprarenal capsules are little bodies of half-moon shape which lie above the kidneys. Their degeneration produces a great lowering of blood pressure. The patient becomes intensely weak, muscular movement becomes impossible, intellectual processes cause great fatigue, and finally blood pressure becomes so low that fatal collapse ensues from lack of blood in the brain. The external symptoms of these cases is a pigmentation, that is, a very dark discolouration of the skin, which develops rather early in the disease. The tongue especially becomes a very dark brown. Areas of pigmentation also occur where the skin is irritated,—at the wrists from the irritation of the coat sleeves, at the edge of the hair from the irritation of the hat. Dr. S. Weir Mitchell, in his Autobiography of a Quack, has described one of these cases very strikingly. The hero of the tale is found dead one morning by the nurse in the hospital, after he has been feeling quite as well as usual for some time.

It must not be forgotten that patients who are burned extensively very frequently die shortly after the accident. A burn that involves more than one-half of the body, no matter how superficial the burning may be, will always have a fatal termination. Deep burns in one part, unless it is some very vital part, are not so serious as extensive superficial burns. Patients with extensive burns frequently remain in encouragingly good condition for several days, and then have a sudden change for the worse. Sometimes death takes place in coma. Sometimes it takes place as the result of a perforation of the duodenum. These perforations of duodenal ulcers may take place as late as a week to ten days after the burn. They are always followed by symptoms of peritonitis and the condition of intense prostration which this brings on. Such cases need to be prepared for the worst after the first acute symptoms of the burn have subsided, when a certain amount of peace of mind is restored.

Cirrhosis of the liver not infrequently causes sudden death. Cirrhosis is an affection in which a large part of the liver substance proper degenerates, and its place is taken by connective tissue. It is typically a disease of people of alcoholic habit. It occurs in those who are engaged in the sale of spirits, though the alcoholic absorption does not take place through the skin, but in a much more direct way. It is most frequent in people who take strong spirits on an empty stomach. Those who are much exposed to changes of temperature are especially liable to form such habits. It is found most frequently in the drivers of wagons and cars, in policemen, and in sea-captains, sailors, and the like. When cirrhosis causes sudden death, it is nearly always by hemorrhage. The hemorrhage takes place from the oesophagus, some of the large veins of which have become dilated until the thin walls are unable to retain the blood. The dilatation is due to interference with the venous circulation in the liver.

Of late years pathologists and medical men, especially those who are interested in children's diseases, have devoted considerable time to the study of certain cases of sudden death, which have long been very mysterious. Infants often die while in apparent good health without any adequate reason that can be found, even on the most careful autopsy. Children of an older growth sometimes die suddenly as the result of some slight shock or fright, or they die after the administration of a few whiffs of chloroform, given to help in the performance of some simple surgical operation, or they die at the beginning of some infectious fever which they ought to be able to withstand without any difficulty. A distinguished pathologist at Vienna, Professor Paltauf, who was the coroner's physician of the city and had a large number of these sudden deaths to investigate, found that in most of the cases one abnormal condition was constantly present. This consisted in an enlargement of the lymph glands all over the body. The lymph glands in the neck were involved, also the tonsils and lymphoid tissue at the back of the throat, the series of lymph glands in the groin, and, finally, there was a hypertrophy of the lymphoid tissue that occurs all along the intestinal tract. This condition of hypertrophy of lymphoid tissue has come to be known as the lymphatic diathesis or constitution. It is nearly always accompanied by a distinct hypertrophy of the thymus gland. The thymus gland is an organ which occurs in the upper part of the thorax of the child, but which atrophies and practically disappears after the age of two years. In these cases it is from twice to three times its normal size in the infant, and in older children it is persistent—that is, retains its primary size, though in the ordinary course of nature it should atrophy. This lymphatic diathesis undoubtedly has considerable to do with the sudden deaths which occur in these patients. What the exact connection is we do not as yet definitely know. Unfortunately, moreover, this lymphatic constitution gives no sure sign of its existence before the occurrence of the fatal termination. Enlargement of the glands of the neck and of the groin, with some enlargement of the tonsils, occurs in delicate children without necessarily being symptoms of the lymphatic diathesis. The enlargement or persistence of the thymus can be better recognised, and doctors now seldom fail to notice it. Where any suspicion of such a condition exists in children of from eight to sixteen or seventeen years of age, proper precautions must be taken to prevent sudden fatal termination of any even mild disease without due preparation. Undoubtedly many of the cases of sudden death under chloroform and ether in children and young persons are due to the existence of this lymphatic diathesis.

Diseases, like tuberculosis and cancer, that run a long but assuredly fatal course, usually terminate unexpectedly. The tuberculous patient particularly will almost surely be planning for next year the day before he dies. This condition of euphoria, that is, of sense of well being, was recognised as associated with tuberculosis as far back as we have any history of the disease. Hippocrates pointed out as one of the symptoms of consumption the spes phthisical or consumptive hope. If the patient has been very much run down, death may take place from thrombosis of some of the arteries. If the thrombosis takes place in the brain, consciousness will be lost, and the patient will often die without recovering it. Patients often develop tubercles in their brain as the result of a spread of the disease beyond the lungs, and then, as a rule, death will take place in the midst of a paralysis, which may be accompanied by loss of consciousness that lasts for several days or a week or more.

Cancer patients also die suddenly, or at least unexpectedly, at the end. Very often in them, as in tuberculosis, thrombosis plays an important rôle in the fatal termination. In cancer of the stomach, peritonitis from perforation of the stomach may close the scene. The fatal termination in cancer of the uterus is often brought about by the development of uraemic symptoms. The new growth in the pelvis involves the ureters, prevents the free egress of urine, and so causes the retention in the system of poisonous substances that should be excreted. Cancer in other parts of the body often causes death by metastatic cancers, that is, offshoots of the original cancer which occur in other organs. Usually these are in the liver, but sometimes they are in the brain, and sometimes in the bones that surround the spinal cord. In the course of their growth they cause pressure symptoms upon the nervous system, and this leads to death. If patients become very much weakened, as is not infrequently the case, thrombosis occurs, and portions of the clots may be shot into the pulmonary veins, and cause death in this way.

Two affections which are quite common, one of them usually involving no danger at all, sometimes cause sudden death. They are varicose veins and a discharging ear. Varicose veins are the enlarged veins which occur on the limbs of a great many elderly people. If these people become run down in health and then exhaust themselves by overwork, the circulation through these enlarged veins is sometimes so impeded that clotting—thrombosis, as it is called—occurs. If a portion of the clot becomes detached, and is carried off into the circulation, a so-called embolus, this may cause sudden death, either by its effect upon the heart, or more usually upon the lungs.

Middle-ear disease causes death, either by producing an abscess of the brain, or by causing thrombosis of some of the large veins within the skull. The dangers involved in a discharge from the ear are now well recognised. Insurance companies refuse to take risks on the lives of persons affected by chronic otitis media, as it is called scientifically. Such persons may run along in perfect good health for years without accident, but a sudden stoppage of the flow may be the signal for the

Certain severe forms of the infectious fevers are very often fatal. These forms are popularly known as black fevers, that is, black measles, black scarlet fever, etc. These fulminant forms occur especially in camps, barracks, orphan asylums, jails, and the like, where the hygienic conditions of the patients have been very poor, and where the resistive vitality has, as a consequence, become greatly lowered. The black spots that occur on such patients are really due to small hemorrhages into the skin. The hemorrhages are caused by a lack of resistance in the blood-vessels and by a change in the constitution of the blood that allows it to escape easily from the vessels. Where such cases occur, patients should be fully prepared for the worst As a rule, the mortality is from 40 to 70 per centum.

Acute pancreatitis is a uniformly fatal disease, though fortunately it is rare. It occurs much more frequently, however, than used to be thought. It occurs in persons over thirty who have been for some years addicted to the use of alcohol. The symptoms of the disease are severe pain in the upper left zone of the abdomen, that is, above and to the left of the umbilicus. This is accompanied by nausea and vomiting. Collapse ensues and death takes place on the second to the fourth day of the affection. This disease may have important medico-legal bearings. Some slight injury in the abdomen, as from a blow or a kick, may precipitate an attack in predisposed individuals. Accusation of murder may result. The mental attitude of the physician and the clergyman with regard to such cases must be very conservative. No opinion as to possible culpability should be ventured.

Cholelithiasis, that is, stone in the bile duct, may not only cause severe pain, but may lead to rupture of the duct and a rapidly fatal termination. Owing to the practice of wearing corsets, gall-stones occur much more commonly in women than in men. Twenty-five per centum of all women over 60 years of age are found to have gall-stones. While these cases suffer from intense pain they are very seldom fatal. But it must not be forgotten that a fatal issue can take place either from collapse and stoppage of the heart, because of the intensity of the pain, or from perforative peritonitis.

The perforation of a gastric ulcer may cause symptoms which rapidly place the patient in a condition in which the administration of the Sacraments is very unsatisfactory. Gastric ulcers occur especially in young women, usually in those who follow some indoor occupation. Its favourite victims are cooks, though laundresses, seamstresses, and even clerks in stores, suffer from it much more than those engaged in other occupations. It occurs by preference in anaemic or chlorotic women. Sometimes, however, as in the case of cooks, the patients may seem to be in good health. Acute pain in the stomach region, followed by symptoms of collapse, should in such persons be a signal for the administration of all the Sacraments. Fatal peritonitis soon brings on a state of painful uneasiness ill adapted to the proper dispositions for the Sacraments.

Two diseases that are fortunately very rare, but which are almost uniformly fatal, deserve to be mentioned here. In both of them the symptoms of the disease are manifested through the nervous system. They are tetanus and hydrophobia. Tetanus occurs as a consequence especially of a wound which has been contaminated by the street dirt of a large city, or the refuse of a farm. It follows deep wounds such as are made by a hayrake or a pitchfork; or seared wounds, such as are made by a toy pistol. A serum for the treatment of the disease has been discovered, but unfortunately the first symptom of tetanus is not the first symptom of the disease, but the preliminary symptom of the terminal stage of the disease, the affection of the nervous system. Practically all cases of acute tetanus terminate fatally. As soon as a patient exhibits the characteristic symptoms, the lockjaw, the stiff neck, and the rigid muscles, all the Sacraments should be administered. In tetanus, as a rule, consciousness is preserved until very late in the disease. In severe cases, however, a convulsive state of intense irritability develops in which the slightest sound or effort brings on a series of spasmodic seizures. Patients must be prepared, then, early in the disease, if possible.

Rabies or hydrophobia is a disease which claims a certain number of victims every year in our large cities. Its symptoms are the occurrence of fever and disquietude, with spasmodic convulsions of the muscles of the throat whenever an attempt is made to swallow. These symptoms come on from three to fifteen days after the bite of a mad dog. Unless the Pasteur treatment has been taken shortly after the bite of the animal was inflicted, no treatment that present-day medicine possesses is able to affect the course of the disease, and patients nearly always die. Their preparation, then, is a matter of necessity as soon as the first assured symptoms of the disease show themselves. [Footnote 4]

[Footnote 4: One cannot help but add a word here as to the cause of the disease, because clergymen can by their advice do something to remedy the evil which lies at the root of the infliction. Hydrophobia is due to stray dogs. In practically every case the fatal bite is inflicted by some animal that no one in the neighbourhood claims. Bites by pet dogs are rarely fatal. If clergymen would use their influence to suppress the dog nuisance we would soon have an end of hydrophobia.]

Alcoholic subjects are very liable to unexpected death from a good many causes. Patients suffering from delirium tremens, for instance, may die suddenly in the midst of a paroxysm of excitement. Such a termination is not frequent, but it has occurred often enough to make it the custom, at asylums for inebriates, to warn friends who bring patients of the liability of such an accident. It is not so apt to happen during a first attack of delirium tremens as during subsequent attacks. It is most frequent among those whose addiction to alcohol for years has caused repeated paroxysms of delirium tremens. The cause of the sudden death is usually heart failure. This term means nothing in itself, but it expresses the fact that a degenerated heart finally refuses to act. Alcoholic poison in the circulation has led to fibroid degeneration of the muscular elements of the heart and made them incapable of proper function, or at least has greatly hampered their action, and the heart ceases to beat.

It must be borne in mind that chronic alcoholism makes a number of serious organic diseases run a latent course. The patient is apt to attribute his symptoms to the after effects of the abuse of alcohol. Unless the doctor who is called in makes a very careful examination, serious kidney disease or even advanced pneumonia may not be discovered. Alcoholic subjects bear pneumonia very badly, and the preliminary symptoms of the disease are often completely concealed by the symptoms due to the patient's alcoholism. Other infectious diseases, as typhoid fever, tuberculosis, and even various forms of meningitis, may run a very insidious course and give but very slight warning of their presence. The result is that these diseases are very frequently fatal in alcoholic subjects.

Old inebriates bear operations badly, and the mortality after any operation in such subjects is distinctly higher than in normal individuals.

One reason for this is that considerably more ether or chloroform is required to produce narcosis in alcoholic subjects than in ordinary individuals. Ether and chloroform are very irritant to the kidneys. The kidneys are prone to be affected more or less in old alcoholic subjects. Death from oedema of the lungs or from some form of pneumonia is not infrequent in these post-operative cases, and gives as a rule but little warning of its approach.

It is clear, then, that alcoholic subjects must be prepared with special care whenever disease is actually present or an operation is to be performed. Too great care can scarcely be exercised in their regard. What would seem overcaution will save many a heartburn to friends and priest, for it is in alcoholic subjects especially that some of the saddest cases of unexpected death without preparation occur.

JAMES J. WALSH.

THE MOMENT OF DEATH

It not infrequently happens that a priest reaches a patient who has just died. Conditional absolution, baptism, or other spiritual ministration might have been offered if there were signs of life, but the heart and lungs are still, "the patient is dead," and the priest leaves the place without doing anything. Yet the patient may not really be dead.

Our knowledge of the precise time the soul leaves the body is very imperfect. There is, we are aware, a close connection between the vital functions of the body, taken together or singly, and cellular activity. If the cells are not destroyed, a vital function sometimes may be restored after its cessation, but if the cells are destroyed up to a certain extent, the vital function is not recoverable. For example, if the various bodily cells of a patient dead from diphtheria are examined microscopically, it will be found that the diphtheria toxin has disintegrated the nuclei of these cells. What number of cells proportionate to the whole in, say, the heart should be destroyed before the vitality of that organ is lost, is not clearly known. Where the cells are intact, or nearly so, mere absence of respiration, or of even the heart movement, are not absolute proof of death. Numerous cases are found in medical records of persons that had been lying under water for many minutes, up to even an hour, but who were restored to life by patient and skilful efforts; and of late remarkable restorations after what was practically death, under anaesthesia and otherwise, have been reported. The technique consists chiefly in rhythmical compression of the heart, commonly after surgical exposure of that organ, with artificial respiration, and, in Crile's method, peripheral resistance is employed to raise the blood pressure. Ludwig in 1842, experimented in cardiac massage, and Professor Schiff at Florence was the first to apply the method to human subjects. Kemp and Gardner, in the New York Medical Journal, May 7, 1904, described various methods used in attempting resuscitation.

Professor W. W. Keen of Philadelphia has collected the records of the chief cases of resuscitation after apparent death (see The Therapeutic Gazette, April, 1904), and some of these are the following: Dr. Christian Igelstrud of Tromsö, Norway, in 1901, was operating upon a woman, 43 years of age, for cancer. During the operation, which was a coeliotomy, she collapsed and her heart ceased beating. After the usual means for resuscitation had been ineffectively tried, her heart was laid bare. Igelstrud took hold of the heart with his hand and made rhythmic pressure upon it. In about one minute the heart began to pulsate. The patient was discharged from the hospital five weeks afterward.

Tuffier (Bull, et mém. soc. de chir., 1898, p. 937) in 1898 had a patient whose heart stopped after an operation for appendicitis. The surgeon had left the operating room, but he returned, laid bare the heart, pressed it rhythmically, and after two minutes it began to move again. The patient breathed regularly, his eyes opened, the dilated pupils contracted, and he turned his head. After the opening over the heart had been closed, however, he died.

Prus (Wiener klin. Woch., no. 21, 1900, p. 486) by the same method started contractions of the heart after 15 minutes in a man that had hanged himself. The effort at resuscitation was made two hours after the suicide had been discovered, but the recovery did not go beyond imperfect movements of the heart, which gradually ceased.

Maag (Centralbl. f. Chir., 1901, p. 20) reports the case of a man who under chloroform anaesthesia ceased breathing and whose heart stopped. After 10 minutes the patient was pulseless, without respiration, cyanotic, and cold. The heart was exposed and compressed rhythmically; it was restored to action, and he began to breathe. He remained alive for 12 hours, seemingly asleep; then he died.

Starling and Lane (Lancet, Nov. 22, 1902, p. 1397) were operating upon a man 65 years of age. The heart and respiration ceased. Lane put his hand into the abdominal incision and squeezed the heart through the diaphragm. After twelve minutes of artificial respiration the lungs and heart began to act. The patient afterward was discharged from the hospital cured.

Sick (Centralblatt f. Chirurgie, Sept. 5, 1903, p. 981) reports a very remarkable case. A boy of 15 years of age died upon the operating table. Three quarters of an hour after the heart had ceased to beat it was laid bare. The flaps did not bleed, the pericardium was bloodless, the heart was motionless, relaxed, and cold. After a quarter of an hour, during which the heart was compressed, and artificial respiration was kept up, that is, one hour after what any physician would call death, the heart was beating and respiration was restored. Two hours later the boy became conscious and complained of great thirst and dyspnoea. He remained in this condition for twenty-seven hours, and during that time his speech was indistinct but intelligible. He then died.

Dr. George W. Crile, of Cleveland, Ohio, reports the case of a woman whose heart movement and respiration had ceased for six minutes. She was restored completely, even without exposing the heart. Dr. Crile uses an inflated rubber suit on the patient to raise the blood pressure by peripheral resistance—he does not expose the heart. He had another case, a man 38 years of age, who "died during operation, was resuscitated, and died again two hours later."

Two Hungarian labourers, whose skulls had been crushed in the same accident, were brought into Dr. Crile's clinic in a dying condition. The heart of one of these men ceased beating as he was brought into the operating room. After nine minutes the surgeons began to work upon him to resuscitate him. They succeeded, but he lived for only 28 minutes.

They then examined the other man and found him dead. Just 45 minutes after this second patient had been brought into the operating room the effort to resuscitate him began. As he had not been observed while the physicians had been engaged with the first man, they do not know when his heart had ceased to beat, but he certainly was dead in the opinion of skilled observers. They resuscitated him so well that he moved his head away from the operator who was relieving the depression of the skull, but he died again in 34 minutes.

These cases are not what is commonly called conditions of suspended animation. All the patients would have been pronounced dead by any physician, and if they had been left untouched, they surely never would have been revived.

There have been about thirty attempts made by surgeons to restore patients who were dead in the full acceptance of the term as used at present. Four of these attempts resulted in complete success, others in a partial recovery, and many were without positive result. The number of complete and partial resuscitations, however, are enough to justify a priest in giving conditional absolution or baptism within an hour, or even two hours, after a patient has to all appearance died, especially in accident cases. We do not know when the soul enters the body, and there is the same doubt as to the moment when the soul leaves the body. In these latter cases we should give the patient the benefit of the doubt.

AUSTIN ÓMALLEY.

THE PRIEST IN INFECTIOUS DISEASES

The subject of infection is complicated, and the medical doctrine concerning it is far from certainty despite the multitude of facts presented by bacteriologists, chemists, pathologists, and clinicians. Before the days of bacteriology the term Infectious commonly was applied to diseases produced by no known or definable influence of any person on another, but wherein common climatic or other widespread conditions were thought to be chiefly instrumental in the diffusion. The contagious disease was one transmitted by contact with the patient, either directly by touch, or indirectly through the use of the same articles.

Now we know that many diseases called infectious are caused by micro-organisms, and we group others under this class because we hold theoretically that they have their origin in microbes not yet isolated. Hence we define an infectious disease as one which is caused by a living pathogenic micro-organism, which enters the tissues from without, and is capable of multiplying therein. These micro-organisms have a time of incubation during which a poison is made in the tissues, and this brings about the intoxication we call the disease.

Infection is a general term that includes contagion; and contagious diseases are infective diseases that may be transmitted directly or indirectly from patient to patient.

The pathological micro-organisms with which we shall deal in this article are (1) the Schizomycetes or Fission-Fungi, which are microscopical organisms that multiply by fission, and are commonly known as Bacteria; and (2) a few Protozoa, which are animal micro-organisms.

The bacteria are classed with plants because, like plants, they derive nourishment from both organic and inorganic material. They have no seeds or flowers, but many of them are reproduced by spores. They consist of cells, single or grouped, which when spherical are called cocci, when rod-shaped, bacilli, when spiral, spirilla. There are various subdivisions of these groups. We do not know whether bacterial cells have nuclei or not.

A micro-organism is a parasite when it can live in animal tissues. It is a saphrophyte when it can exist outside animal tissues. If a parasite cannot exist outside animal tissues, it is an obligatory parasite; if it can, it is a facultative saphrophyte. Similarly the saphrophytes are classed as obligatory saphrophytes and facultative parasites. Pathological micro-organisms have very complicated products which are in large part poisonous.

Bacteriologists require seven conditions to prove a micro-organism the specific cause of a given disease, and all these conditions have been fulfilled for anthrax, diphtheria, and tetanus. The specificity has been satisfactorily settled for glanders, malaria, tuberculosis, actinomycosis, gonorrhoea, and malignant oedema. It has been practically settled for typhoid, influenza, the Madura disease, and the bubonic plague; and incompletely defined for leprosy, relapsing fever, and Malta fever.

There are certain diseases which are not called specific, because they may be produced by various micro-organisms. These are pneumonia, osteomyelitis, septicaemia, pymaeia, endocarditis, meningitis, erysipelas, angina Ludovici, broncho-pneumonia, and similar maladies. Cholera and dysentery also might be grouped with these, as cholera appears to be produced by various vibrios and dysentery by different amoebae.

There are other infective diseases, in which we have not yet found the causative micro-organism, but we presume its existence. These are: rabies, syphilis, yellow fever, dengue, typhus, mumps, whooping-cough, smallpox, measles, scarlet fever, and others among the exanthemata.

Malaria and similar diseases are caused by plasmodia, which are protozoa and not bacteria.

The priest is almost as frequently exposed to the danger arising from contagion as the physician is, and a priest that often ministers to the sick is liable to grow imprudently indifferent to danger. For one priest that is too much afraid of disease we find a hundred that have not sufficient dread.

No matter what medical science may say to the contrary, many priests hold that they have often left smallpox cases, for example, without disinfecting themselves, and that they have not spread the disease. This is a very rash assertion. It is absolutely certain that smallpox has been communicated to susceptible persons by those coming from patients ill with that disease merely passing the susceptible man on the street. The number of persons that will not take smallpox when exposed to it is very large. In Washington in 1895, during an epidemic of smallpox, 187 persons, to my personal knowledge, were exposed to one group of 39 smallpox patients without taking the disease. The unharmed had been present in sick-rooms or had even nursed the patients, not knowing that the disease was smallpox. In this epidemic eight persons lived in the same rooms with, or visited frequently, two patients that afterward died of virulent smallpox, and none of the eight took the disease. One of these eight, however, went into a dramshop, had one glass of beer and left immediately, and in fourteen days afterward (the average time of incubation) we took the barkeeper to the smallpox hospital. This barkeeper had not been exposed to smallpox except by contact with the man mentioned here. There were about 60 cases of smallpox in that epidemic, and we traced every one to direct or indirect contact with one initial case.

If we were infected by every exposure to contagious disease the world would be depopulated. It is true that you cannot give some persons diphtheria if you actually put the Klebs-Loeffler bacillus into their mouths, and nurses and physicians in consumptive wards have the tubercle bacillus in their nostrils without ill effect. So for many diseases; but it unfortunately remains true that there are susceptible persons everywhere who will at once take a disease when they are exposed to it.

Immunity changes in the same person. Starvation, fatigue, loss of blood, unsuitable diet, exposure to heat, cold, and moisture, and other influences lessen the power of resistance to infection. Men vary almost as do the lower animals as regards infection. The quantity of tetanus toxin that will kill 400 horses will not bother a hen; Algerian sheep and the white rat are not affected by anthrax, but other sheep and the brown rat are very susceptible; a hog will not take glanders, man and a horse will; men, cattle, and monkeys have tuberculosis, dogs and goats do not; white men with few exceptions are susceptible to yellow fever and malaria, negroes are practically immune; negroes readily succumb to the fatal sleeping sickness, white men are almost immune; and similar differences are observable in the same race or family.

The question of immunity to infectious disease is very difficult to make clear because it is so technical, and it is only a theory at best. The poison of an infectious disease kills by splitting and destroying the nuclei of the body's cells. The toxic products of the micro-organisms seem to become chemically united with certain molecules of the body cells and to inhibit the normal function of these molecules. According to Erlich's theory there are other molecules in cells which neutralise toxic molecules, and when the neutralising molecules appear in excess the patient recovers. These neutralising bodies are called antitoxins.

Some antitoxins are always present in cells, and where the normal quantity of these is used up in neutralising toxins, other antitoxic bodies are formed, until finally the excess of these is thrown off into the blood serum. After they are called into being by the excitation of some toxic products, like those of the typhoid bacillus for example, the antitoxins remain in the blood for years, ready to neutralise at once any influx of fresh infection. In other diseases, like diphtheria and pneumonia, they are soon lost,—hence the recurrence of such diseases. The acquired antitoxin lasts after smallpox, vaccinia, yellow fever, scarlet fever, measles, typhoid, mumps, and whooping-cough; it is very transient after pneumonia, influenza, diphtheria, erysipelas, and cholera.

In serum therapy antitoxins are artificially excited into being in the blood of beasts. This artificially prepared antitoxin is injected into the blood of, say, a diphtheria patient, and the poison is at once neutralised, instead of leaving the patient to make his own antitoxin and letting him perhaps fail in the effort.

The antitoxin produced in the contest of the body cells against some diseases will not only neutralise the toxin of a particular disease, but it will also neutralise the toxin of a second disease. By vaccinating a person we inoculate him with vaccinia or cowpox. His body cells make an antitoxin which neutralises the toxin or virus of cowpox, he recovers from this light disease, and the antitoxin now remaining in his body prevents for years another successful inoculation with cowpox. It does more: in 90 per centum of cases it will prevent successful infection with smallpox.

Smallpox (the pocks, pokes or pockets of matter,—opposed to the great pox or syphilis) has been known from very early times—probably even from 1200 B.C. The name "small pokkes" was first used in England in 1518. The disease was brought to America in 1507.

It may be communicated from the sick to the healthy (1) by persons suffering with the disease; (2) by bodies of persons that have died of smallpox; (3) by infected articles; (4) by healthy third persons; (5) by the air, to persons living even at some distance; (6) by inoculation. The poison enters the body by the mucous membrane of the nose, mouth, or respiratory tract, and probably through the mucous membrane of the stomach and through the broken skin.

Patients can communicate the disease probably during the period of incubation (from 5 to 20 days after exposure to the disease—commonly about 14 days); and certainly from the initial stage until no trace is left of the final skin-desquamation. The infection is most active during the formation and duration of the pocks. The mildest smallpox in one person can cause malignant smallpox in another, and vice versa. The mortality in the unvaccinated is between 40 and 50 per centum.

A typical case of confluent smallpox at its height is the ugliest disease in appearance and stench and almost in substance, known to medicine. Anyone liable to infection by it, or likely to carry it to others, who says he is "not afraid of it," has either never seen it and he is talking childish nonsense, or he has seen it and he is a fool.

The face is a bloated mass of corruption; the eyes are swollen shut; the nose, cheeks, lips, and neck are puffed out enormously; the mouth is a large sore, ulcerous, and spittle trickles from it ceaselessly. The fever is up to 103 or 105 degrees; there is an unquenchable thirst, a vile stench, sleeplessness;

often delirium is the only relief, and there is one chance in two of a disfigured recovery. Tobacco, alcoholic liquor and a walk in the fresh air will not disinfect the visitor to such a disease. Years ago I investigated in the laboratory the popular notion that tobacco is a disinfectant. I found that bacteria, the diphtheria bacillus and swarms of others more delicate, will grow as well in the presence of a large piece of "Navy Plug," as when tobacco is absent. Chewing tobacco, whiskey, a walk in the fresh air as disinfectants, the Sioux medicine-man's powwow, the hind leg of a rabbit as a charm, are all in the same category.

The first and chief protection against smallpox is vaccination. Vaccination does not always prevent infection by smallpox, but it does prevent it in more than 90 per centum of exposures to the disease. Welch reported in 1894 that the death-rate in one series of 5,000 cases of smallpox was 58 per centum in the unvaccinated, and 16 per centum in the vaccinated, but the vaccinated took the disease in less than 10 per centum of the exposures. During the Franco-German War in 1870-1871, the Germans who had a million vaccinated men lost 458 soldiers from smallpox while a great epidemic of smallpox was existing in Germany; the French, who were indifferent to vaccination, during the same time lost 23,400 men from this disease alone. In the United States, where there is no compulsory vaccination except such attempts as school boards make, there were between July and December, 1903, 13,739 cases of smallpox; in Germany, where there is a compulsory vaccination law, there was no smallpox at all, during the same time, except 14 cases in two seaports, Bremen and Kiel, whither the infection had been brought from without.

Before 1874 there had been no compulsory vaccination law in Germany except for the army. In 1871, 143,000 Germans died of smallpox. Since the law went into effect in 1874 the disease has been stamped out, until there was between July and December, 1903, only one death from smallpox in Germany.

The chart on page 175 will show very graphically the effect of vaccination upon smallpox.

In October, 1898, smallpox was endemic in Puerto Rico; in December, 1898, it was epidemic; in January, 1899, it was all over the island and spreading rapidly. In February, 1899, compulsory vaccination was begun and carried out for only four months, when 860,000 vaccinations had been made in a population of about 960,000 people. The death-rate from smallpox dropped from 621 a year to 2.

During the century preceding Jenner's discovery of vaccination, according to Neimeyer's calculation 400,000 people died of smallpox each year in Europe. Bernouilli, a trustworthy statistician, says that during that same century, "Fully two-thirds of all children born in Europe were, sooner or later, attacked by smallpox, and on an average one-twelfth of all children born succumbed to the disease."

Early in the sixteenth century 3,500,000 people in Mexico had smallpox (Prescott's Conquest of Mexico). In 1707, in Iceland, 18,000 of the population of 50,000 died of smallpox; and in 1891, 25,000 persons in Guatemala died of this disease. In 1875 there were anti-vaccination riots in Montreal, and as a consequence most of the younger inhabitants of that city were not vaccinated. In 1885, smallpox was brought in from Chicago; 3,164 persons died of the disease; of these 2,717 were children under ten years of age, and thousands had the disease.

Vaccination may render one immune to smallpox for many years, but if the disease is epidemic it is well to renew the vaccination after about eight years. In normal vaccination, where the lymph has been derived from a reliable source, on the third or fourth day pale red papules develop at the point of inoculation, and about the tenth day these have become pustules. The vesicles dry gradually, and between the fourteenth and twentieth days the scab falls off, leaving a pitted scar. About the fifth day an aureola of inflammation forms around the pocks, from a quarter of an inch to two inches in extent, and the inflamed area may be somewhat sore. A shield should be kept over the vaccination spot for two days, and this is then to be replaced by a piece of sterile gauze held in place by narrow strips of sticking-plaster above and below the inflamed area. Sometimes hives and other rashes occur in vaccination, but they are unimportant.

Where there is a very sore arm or other trouble, the cause may be a pre-existing unhealthy condition, like scrofula for example, or the patient has scratched the pocks, or infected them from his clothing, or the vaccine lymph was unsterile. A careless and dirty vaccinator might infect an arm with pus organisms. If good glycerinated lymph, not too fresh or too old, is used, there is seldom any trouble; but in any case all the annoyance that may come from vaccination is infinitesimal when compared with the smallpox it averts.

We may take a smallpox case as a typical contagious disease in which the priest is to give the last Sacraments; and the disinfection and other precautions observed in such a visit will serve for any other very contagious disease. For only typhus and one or two other maladies are the precautions so elaborate as those needed in smallpox.

There is a dress, called "Dr. Hawes' Antiseptic Suit," and in time of epidemics a priest should have one of these suits, or one made after it as a pattern—they can be obtained in the shops for two or three dollars. They cover the entire person, even the shoes, and they make unnecessary the changing of clothing and the disinfection of the exposed parts of the body. The hands of the priest may be left bare after fastening the sleeves of the suit about the wrists, or he may wear surgeon's thin rubber gloves. In visiting a patient that has any of the contagious diseases mentioned in this chapter, the priest should never touch his own face with his hands after he has entered the sick-room until he has washed them in a bichloride of mercury solution.

A ritual should not be taken into a smallpox room, because a book cannot be disinfected without rendering it useless. The priest should memorise the prayers and ceremonial, or write them out on paper which can be burned in the hospital or the patient's house.

The priest may be obliged to administer baptism, to hear confession, to give the Viaticum and Extreme Unction. Before going to visit a smallpox patient let him find out from the physician in attendance whether the patient can receive the Viaticum, whether he can swallow it or not, whether he can open his mouth enough to take it. Ask also about the possibility of vomiting. Only a very small particle is to be brought in the pyx.

The leather cover for the pyx should not be taken into a smallpox room. Set the pyx inside a corporal, wrap the corporal in paper, and put this package into the pocket of the Hawes suit before entering the room.

As to the use of a stole,—the moralists say "graviter peccatur ab eo qui sine urgente necessitate sine ulla sacra veste unctionem administrat." There is a grave necessity here for doing away with the stole because of the difficulty in disinfecting it, unless you have one made that can be put into boiling water for ten minutes before you leave the patient's house.

The oil-stocks should contain only as much oil as is necessary for the single occasion, because what remains, with the cotton, should be burned in the patient's house.

Do not remain in the room longer than you must unless you have had smallpox. If there is any prayer or ceremonial that can be omitted, by all means leave it out. Lehmkuhl says that the penitential psalms and the litanies may be omitted. Baptise by the short form.

St. Alphonsus Liguori (Theol. Mor., lib. 5, tr. 5, n. 710) tells us there is no obligation to anoint both eyes and both ears, "si adsit periculum infectionis," but danger of infection is not materially increased by anointing both sides. Lehmkuhl adds, "excepta dispensatione Sedis Apostolicae addatur unctio pedum." When the feet are to be anointed do not touch the bed-clothing,—tell the nurse to uncover the feet.

St. Alphonsus (loc. cit., n. 729) speaking of extreme unction has these words: "Pastor ratione officii tenetur sub mortali dare lis qui petunt, nisi justa causa excuset: etiam tempore pestis, modo possit absque periculo vitae; cum eo non teneri docent Tann. Dian.," etc. If you have not had smallpox you certainly risk your life by going into the room of a smallpox patient, and the danger of infection is greater in typhus; but suppose a pastor were inclined to take advantage of the excuse, he would be obliged at any risk to go into such a room to hear confession or to baptise, and if he hears confession he may as well stay for the anointing.

If you anoint a patient that has confluent smallpox you probably can not wipe away the oil, because the skin will be pustular. Wipe the oil-stock carefully; then all cotton used should be wrapped in paper and burned in the paper before you leave the house. After anointing, you had better wash your hands carefully in water in which a bichloride of mercury tablet has been dissolved—do not use soap and do not put the bichloride in a metal vessel. Wash your hands thus before you leave the sick-room.

If the patient can receive the Viaticum let him lie on his back, and you should drop the Host into his mouth without touching him with your hand. St. Alphonsus says: "non licet tempore pestis porrigere Eucharistiam medio aliquo instrumento … sed manu danda est" There is no need of an instrument. If there are any crumbs left in the pyx make the patient take them. St. Alphonsus says this may be done, and it would be almost certain infection to take them yourself if you have not had smallpox recently. Let as little ablution water as possible be given to the patient.

When you leave the room, put the pyx, oil-stocks, corporal, and stole in a pan of water and boil them for ten minutes. This will disinfect them thoroughly and will not injure them in any way. Then take off the Hawes suit as near the street-door as possible and wet it with bichloride solution. Wash your hands again in the bichloride solution and rinse off the bichloride; take the pyx, oil-stocks, corporal, and stole and leave immediately. Do not touch the door-knob when going out—let some one open the door for you—and do not shake hands with any one.

Typhus fever is now rare in America, but there was an outbreak in New York City in 1881. This was the fever that killed multitudes of Irish emigrants about the middle of the nineteenth century. It is called also spotted fever, camp, jail, ship, and hospital fever, and it has many other names. The name typhus is from , a smoke or fog, and it indicates the befogged, stuporous condition of the patient. Typhoid fever is so called because it has some resemblance to typhus.

The specific cause of typhus is unknown, but the contagion develops and reproduces itself in the body of the patient. It is thought that the contagion exists in the secretions and excretions of the body and in the exhalations from the lungs and skin. The infection can certainly be carried by clothing, dust, furniture, conveyances of all kinds, and dead bodies, and it remains active for months. It may be transmitted through the air for short distances, not nearly so far as the air will carry the contagion of smallpox. In well-ventilated rooms there is less danger of infection, and a typhus patient should have at least 1,500 cubic feet of air space. The contagion may be transmitted in all stages of the disease and during convalescence.

Physical weakness, anxiety and worry, improper food, and poverty, are disposing conditions for infection by typhus. The mortality is about 10 per centum—much less than that of smallpox.

In giving the last Sacraments to a typhus patient exactly the same method should be followed as that observed for a smallpox patient. Keep as far from the patient as possible. After you touch him in anointing or in giving other Sacraments step away from him to say the necessary words. Do not stand between him and an open fireplace, window, door, or ventilator.

Relapsing fever, or famine fever, caused by Obermeier's spirillum, is sometimes associated with typhus. It has a mortality that can go up to 14 per centum in unfavourable circumstances, but the disease is not more contagious than typhoid under hygienic surroundings. Wash the hands in bichloride solution after visiting a case, and do not touch the door-knob or things in the room.

Rabies (called also hydrophobia in man) is a rare disease. It is communicable by inoculation, but it is very doubtful that the disease has been communicated from man to man. The saliva from a person suffering with rabies if injected into a warm-blooded animal will cause rabies, and on that account it is prudent to use care in touching such a patient in administering the last Sacraments. The virus might enter through an abrasion on the priest's hand.

There is a false hydrophobia observed in excitable persons that have been bitten by a dog thought to be mad. The dog that has genuine rabies grows sullen, it hides in corners, and it snaps at everything presented to it A sticky, frothy mucus drivels from its mouth and its eyes become red. It will run straight ahead, snapping at anything it meets; it swallows small stones, chips, and similar objects; it does not avoid water. It howls, grows lean, and its hind legs and lower jaw become paralysed.

In man there is a premonitory stage; a furious stage, which lasts from about a day to three days; then a final paralytic stage. It is well to wait for the paralytic stage before anointing the patient, because in the other stages the slightest touch causes violent spasms. Confessors should note that the virus of rabies excites the sexual centres.

Scarlatina or scarlet fever first appeared in North America in Massachusetts in 1735. It is especially an April disease here. One attack commonly makes the person immune for life. It is a disease of children, but it attacks adults, and it is fatal among children old enough to receive the last Sacraments. Some epidemics are very malignant; and in such times all the precautions mentioned in speaking of the visitation of smallpox patients should be observed. The contagion is spread just as that of smallpox is spread, except that it is not carried through the air so far.

Diphtheria is a disease of children, but it also can be fatal to adults and to children old enough to receive the last Sacraments. It is caused by the Klebs-Loeffler bacillus, and it most frequently attacks the throat and nostrils. It can start in a cut in the skin, or on any mucous surface, as the inside of the eyelid. The contagion is not in the breath, but it can be coughed out. It is in the saliva of the patient and it gets on his hands and on what he and the nurse touch. It is not nearly so infectious as smallpox and scarlet fever.

In visiting such a patient the priest should be careful not to touch anything in the room, and he should wash his hands in the bichloride solution after a visit. He must also wet the soles of his shoes with the solution. He should be very careful lest a child suddenly cough fine sputum containing the bacillus into his eyes. Diphtheria in the eyes would destroy sight, and I have seen a pair of spectacles save a man in a case like that. A detailed description of the disinfection in diphtheria is given in the chapter on Infectious Diseases in Schools.

Glanders is sometimes transmitted from beasts to man, and it is almost always fatal in the human subject. The disease is caused by the glanders bacillus. Horses, asses, dogs, cats, goats, and sheep are susceptible to the disease; pigs are somewhat susceptible; cattle and birds are immune. The infection is in the discharge from the nose of the patient and on the skin eruptions. The same precautions are to be taken as are needed in a diphtheria case.

Influenza, called popularly the grippe, is caused by the bacillus influenzae, which was isolated by Pfeiffer in 1891. The bacillus is found in the nasal secretions and in sputum; it dies in from twelve to twenty-four hours when dried. The disease is contagious, and it is often fatal in alcoholics, the overworked and harassed, and in those that have chronic diseases. In any case it is a serious malady. Disinfect the hands after visiting a case.

Dengue becomes epidemic at times, especially in the Southern States. The disease is very severe, painful, and depressant, but the mortality is quite low except in complication with other maladies. Its cause is not known. It is very contagious and has symptoms which belong to the class of disease in which are scarlatina and measles. The priest should act as in a case of scarlatina.

There is a form of pneumonia which spreads so widely and rapidly that it is called epidemic pneumonia. In visiting patients afflicted with this disease the priest should act as in a diphtheria case.

Epidemic cerebrospinal meningitis is a very fatal disease at times in America. Even those patients that survive are frequently made blind or deaf, or are left injured otherwise. The malignant type is nearly always fatal. In some epidemics the mortality is as high as 75 per centum. The visiting priest should act as in a case of diphtheria, although the danger of direct infection is not great.

Tuberculosis is a chronic febrile disease, caused by the bacillus tuberculosis, a parasitic micro-organism discovered by Koch in 1882. One-seventh of mankind die by this disease. The bacillus remains virulent a long time after it leaves the human body, but it is soon killed by sunlight.

Tuberculosis of the lungs is spread especially through sputum. In the room occupied by the patient, the clothes, furniture, walls, doors, and floor are infected by the bacilli coughed out, even when the consumptive is careful to disinfect the sputum, and, by the way, he rarely is careful. When the priest visits a consumptive's room he should disinfect his hands with bichloride.

Leprosy is caused by the lepra bacillus, discovered by Hansen in 1871. It is present in many parts of the body, especially in the glands and nervous tissues, and it is found in the mucosa of the mouth and in the nasal secretions. It is very profusely distributed in the corium of the skin. The name comes from , scaly.

Leprosy is present here and there along the Mississippi valley from Minnesota and Wisconsin to Louisiana. It is found also in California, Florida, and the Dakotas, in the Philippines, the West Indies, and the worst infected part of the world is the Hawaiian Islands.

The bacillus has not been found in rooms used by lepers, nor in the soil of their graves. Inoculation by leprous material has failed so far undoubtedly to cause leprosy. There is much dispute concerning the contagiousness of this disease. The Dominican Sisters nursing in the Trinidad asylum have been in constant contact with the lepers for about thirty years but none of them has yet contracted the disease. Zambaco Pasha tells of a family which has lived in the leper asylum at Constantinople for three generations and no one in the family has been infected. Father Damien, however, in Molokai, and Father Boglioli, in New Orleans, did contract the disease. There have been cases of infection from man to man, but ordinarily it seems that some unknown factor must be present to insure infection.

A priest need have no more fear in visiting a case of leprosy than he should have in visiting a case of tuberculosis—not so much. He may wash his hands in bichloride solution after anointing a leper, but it is scarcely necessary to do even that.

Actinomycosis (, ray-fungus) is a disease caused by actinomyces, a micro-organism that partly resembles a bacterium and partly a fungus. The disease can be fatal. It is very improbable that it ever passes from man to man, but as a matter of prudence the priest should wash his hands in bichloride after anointing such a patient.

Septicaemia, or blood-poisoning, can be brought about by different pyogenetic bacteria,—the varieties of the staphylococci (irregularly grouped cocci), streptococci (chain-cocci), pneumococci, and others. The danger of infection is so slight that it may be neglected.

Erysipelas can be fatal, especially in alcoholics, the aged, and in chronic diseases. Erysipelas is contagious, especially if the bacteria get into an abrasion in the skin. Patients having this disease sometimes grow delirious and violent, and the priest should be careful how he handles them. Disinfect the hands after anointing such a patient.

Tetanus, or lockjaw, is not communicable except by inoculation. The bacillus, which was isolated by Kitasato, the Japanese bacteriologist, in 1889, is found everywhere in soil, hay dust, floors, on old nails, especially on the floors of old wooden slaughter-houses. It grows best in deep wounds where it is shut off from the oxygen of the air. Hence the danger of treading upon a nail that has been lying near the ground.

Beriberi, a disease observed especially among seamen, appears at times in our coast towns. It is always a very serious malady and sometimes it is rapidly fatal. The infective agent, which is not known, is not undoubtedly communicable from man to man, but it is carried from place to place, and it clings to ships and buildings; it thrives in hot, moist, crowded places. The priest should disinfect his hands after visiting a case.

Anthrax, called also wool-sorter's disease and splenic fever, is a very fatal disease, and the bacillus is communicable to any one through an abrasion of the skin, through the intestines by swallowing it, or through the lungs by breathing it in in dust. Disinfect the hands and the shoes after visiting a patient. Be careful not to touch anything in his room.

The bacteria that cause typhoid fever, Asiatic cholera (which has been epidemic in America) and epidemic dysentery must get on the hands, or on food, or in water, and thus reach the mouth and be swallowed before they produce these diseases. Act in cholera as in anthrax, and disinfect the hands after visiting a case of typhoid.

The bubonic plague, the most fatal of all epidemic diseases, has already appeared in California and Mexico. It is caused by a specific bacillus isolated by Kitasato and Yersin in 1894. The disease is communicated by contact and it is seemingly also miasmatic.

The terrible plague of the Black Death that swept over Europe from 1347 to 1350 was a malignant form of the bubonic plague. Over 1,200,000 people died in Germany, and Italy suffered much more. In Vienna for some time about 1000 people a day died and were buried in great trenches. Venice lost 100,000 inhabitants, and London lost more than that. In both Padua and Florence only one-third of the inhabitants were left alive; at Avignon the Rhone was consecrated so that bodies might be thrown into it for burial; and ships drifted about the coasts of Europe with dead crews. Hecker, in his study of this plague, says that nearly one-fourth of the population of Europe died in that visitation. Civilisation was wellnigh overthrown in the panic. In Germany, Italy, and France the Jews were accused of poisoning the wells and thus causing the plague, and they were slaughtered by thousands. At Strasburg 2000 Jews were burned to death in one holocaust; at other places, as at Eslingen, in despair the Jews set fire to their synagogues and destroyed themselves. The Great Plague of London in 1665, in which 70,000 persons died, was also the bubonic plague.

The mortality is about 90 per centum in some epidemics. The bacillus leaves the body in the faeces, flies carry it to food, it thus gets to rats and mice, and it is carried from place to place. Rats, however, are commonly infected as if by a miasm before the disease appears in man. There is dispute as to the communicability of the plague from man to man by contact with fomites, but it is practically certain the disease can be thus transmitted. Kitasato once succeeded in producing the disease in animals by inoculation with dust taken in an infected house. Merely touching a patient does not apparently convey infection, yet some authorities hold that in time of epidemic the contagion is transmitted even through the air, especially on the ground floor of houses. Perhaps mosquitoes are the medium of infection, as they are inclined to fly low.

In visiting a case of bubonic plague the priest should be as cautious as if he were attending a smallpox patient. After death by smallpox, plague, typhus, cholera, scarlatina, diphtheria, and measles the funerals should be private and the bodies should not be taken to the church.

Malta Fever, or bilious remittent fever, is found in some of the islands taken from Spain. It has a low mortality and is not contagious. Bruce in 1887 isolated the bacterium that causes it.

We do not know the cause of yellow fever despite the claims of Sanarelli that he has isolated the specific micro-organism. Recently American physicians discovered that it is transmitted from man to man by mosquitoes that belong to the genus Stegomyia, the Stegomyia Fasciata especially. If a yellow fever patient is put into a room in which the mosquitoes have been killed and the doors and windows are screened, he is as harmless, as far as contagion is concerned, as a man with a broken leg. The disease is not spread by fomites.

Malaria is caused by plasmodia, which are protozoa, not bacteria, and it is carried from case to case by mosquitoes of the genus Anopheles. So certain are we that this is the mode of infection that the expression "no anopheles, no malaria" has almost become a medical axiom. A bite from an anopheles mosquito does not cause malaria unless the particular mosquito has previously bitten a malaria patient.

The stegomyia flies and bites in the early afternoon and again at night, the anopheles flies and bites after sunset. In visiting a case of pernicious malaria or one of yellow fever avoid the bites of mosquitoes by gloves and a piece of netting, and there is no danger whatever.

The stegomyia mosquitoes are tropical and subtropical, but they can live as far north as Philadelphia and even farther. The anopheles is especially a northern insect. The ordinary culex mosquito, when it alights upon a wall, stands with its body parallel to the wall, as a house-fly stands; the anopheles mosquito stands with its tail raised from the wall at an angle. A mosquito lays its eggs in any pool of still water, and the "wrigglers" seen in an open rain-barrel are the larvae from these eggs. The larvae come to the surface of the water to get air, and they may be smothered with petroleum; but the only effective way to get rid of malaria and yellow fever is to drain or fill pools of water and marshes. Mosquitoes will breed also in the small still bights along the edges of running streams; in old tomato cans that contain rain water; in any still water, fresh or salt.

AUSTIN ÓMALLEY.

INFECTIOUS DISEASES IN SCHOOLS

Cases of diphtheria, scarlet fever, measles, and even smallpox are not seldom found in schoolrooms, and much anxiety can be averted and the spread of infection can be wholly or in great part averted by a knowledge of disinfection.

The laity will often follow the advice of a priest in matters of hygiene when they are inclined to rebel against the regulations of health departments and the suggestions of physicians, therefore a preliminary explanation of methods for the prevention of infection in the family will be advantageous; prevention in the family is also intimately connected with prevention in the school. Methods useful in the family are useful also in convents and boarding-schools.

As regards diphtheria, the chief causes of the spread of this disease are mistaken diagnosis, imperfect isolation, incomplete disinfection, and, paradoxical though it may seem, a lack of susceptibility to the disease in a large number of children.

Many physicians are still under the grave error that diphtheria can always be recognised without the aid of the microscope, and that membranous croup commonly kills. All scientific writers upon diphtheria agree that it is caused by the Klebs-Loeffler bacillus. They also hold that there is a disease called membranous croup, as distinct from diphtheria as typhoid is, but that membranous croup is a comparatively harmless and non-contagious disease. Two per centum is a liberal mortality in membranous croup, yet a certain class of physicians are constantly reporting deaths from this disease. In a series of 286 cases (not deaths) diagnosed as membranous croup by physicians of New York City a few years ago, Park found the diphtheria bacillus in 229, or 80 per centum. I have never examined the throat of a child dead from so-called membranous croup in which I did not find the diphtheria bacillus. This is the experience of almost every bacteriologist who has had to do with diphtheria. Some men report deaths from diphtheria as thrush! These deaths might just as truthfully be attributed to the wearing of linen collars.

On the other hand, according to Baginsky of Berlin, Martin of Paris, Park of New York, and Morse of Boston, from 20 to 50 per centum of the cases admitted even to diphtheria hospitals have not diphtheria at all. Bacteriologists find that about 35 per centum of the cases reported by physicians to be diphtheria are really nothing but tonsilitis or pharyngitis, with now and then a case of membranous croup. Without a bacteriological diagnosis, therefore, 35 families in each 100 quarantined (where quarantine laws exist) are unjustly quarantined and subjected to the trouble and expense of useless disinfection. The suffering this can cause to a poor family, whose small business is often ruined by quarantine, is a matter for very serious consideration. Again, no matter what experience a physician may have had, he can not in many cases differentiate diphtheria in its early stages, or in children of good resisting power, from comparatively harmless throat affections. The extraordinary resisting power against diphtheria shown by some children and adults has been described by Wassermann (Zeitschrift f. Hyg., 19 B., 3 H.). He found one series of 17 children, from one and a half to eleven years of age, and 34 adults, in which 11 children and 28 adults were not only immune to diphtheria, but some of them had enough antitoxin in their blood to neutralise a tenfold fatal dose of diphtheria toxin. This explains many mysterious outbreaks of diphtheria: such immune persons are infected and they carry about the disease unconsciously because they are not ill themselves. I have seen a mother kiss a child dying of malignant diphtheria and the woman did not get even a sore throat, but I know of another case exactly like this in which the mother died from the infection.

There are bad cases of diphtheria which the experienced physician can diagnose as soon as he enters the patient's room without even looking at the throat, but the lighter cases that are dangerous are not easily recognised. I have seen two children of a family in Washington attacked with a slight throat soreness after one child had died of diphtheria in the house. The cases of these two children would never even suggest diphtheria if that first child had not had the disease. Both these patients died within ten days of syncope without the formation of any membrane, but the diphtheria bacillus was present microscopically. To the moment of death there was nothing in the symptoms of these two children to show diphtheria to the naked eye. From a personal experience with more than 800 cases of diphtheria in hospitals and as a medical inspector, I feel certain that light attacks of diphtheria can not be diagnosed without the aid of the microscope.

The immunity mentioned above explains the fact that the Klebs-

ESSAYS IN PASTORAL MEDICINE

Loeffler bacillus is sometimes found in healthy throats, and the person that has such a throat is really more dangerous than a patient that is ill with diphtheria, because we cannot guard ourselves against him. School-children at times have what appears to be mere sore throat but which is really diphtheria in the naturally immune.

All cases of sore throat in school-children should be examined bacteriologically, but unfortunately the bacteriological examination for diphtheria is a complicated process which requires an expert bacteriologist and a laboratory. The cost of a laboratory fitted for this diagnosis alone is not great, but it is not easy to persuade small city governments that they need such plants.

The only resource, then, is to treat every suspicious case of sore throat as if the disease were really diphtheria, until a diagnosis is established as near the truth as possible. Children that are afflicted with throat inflammations should be kept from school. The people should be taught the necessity of isolation and disinfection; they should be warned against patent disinfectants, and told to ask competent physicians to advise them in disinfection.

Diphtheria is not directly caused by unhygienic surroundings. A disregard for hygiene disposes a child for infection if the child is exposed to the bacillus. The specific germ must be introduced into the patient's mouth or nostrils. When a child is infected with diphtheria the breath is not a medium of contagion. The sputum, spat out or coughed out, is a means whereby the disease is spread. The bacillus is in the patient's mouth and nostrils; it gets upon his hands by contact, upon eating utensils, upon whatever touches the mouth of the sick person. The bacillus does not float in the air of even the sick-room, except in those cases where dried sputum is stirred up by sweeping or attrition of other kinds.

In a boarding-school or family when a diphtheria patient is found, select a room set off as far as possible from the rooms commonly used, and before putting the patient into this room remove all curtains, upholstered furniture and carpets from it that are not so cheap or so worn that they may be destroyed after the patient's convalescence, or which are of such texture that they will not be destroyed by water or disinfection by heat. In any case the less there is in the room the easier the disinfection will be.

Use the mattress upon which the patient had slept before you discovered the nature of the disease. Books should be removed, because an infected book can not be disinfected except upon the outside. The room is not to be swept while the patient is in it,—dust may be wiped up with a damp cloth. The cloth is to be disinfected before it is sent out of the room.

The popular notions regarding sulphur as a disinfectant after diphtheria are erroneous. Sulphur fumes in certain definite quantities will disinfect after smallpox, scarlet fever, measles, and some other diseases; these fumes will also kill the diphtheria bacillus, if the bacillus is wet and exposed directly; but if it is buried in sputum or in clothing the fumes will have no effect whatever upon it. The disinfectants to use are acid bichloride of mercury and heat. Formaldehyde does not penetrate well enough to be reliable in diphtheria.

When the patient is taken to the room prepared, let a mixture of one ounce of bichloride of mercury in the powdered form, in two ounces of common hydrochloric acid (not the dilute hydrochloric acid used in medicine), be obtained. This is a violent poison, and it must be kept out of the reach of children and careless persons. Two teaspoonfuls of this solution in an ordinary wooden bucket filled with water to within two inches of the rim makes the disinfecting mixture. A wooden washtub nearly filled with this disinfectant, mixed in the bucket as directed, should be kept near the door of the room, and all towels, sheets, and soiled linen must be soaked in this tub for twenty-four hours. After that any one may handle these articles with perfect safety. The articles that have been soaked for twenty-four hours should be rinsed in ordinary water to remove the acid, and they may then be washed. The nurse should not touch the outside of the tub with infected articles while putting these in the disinfectant. Do not make the disinfectant stronger than directed here, or it will destroy the articles soaked in it, and for the same reason do not leave them in it longer than twenty-four hours.

If the attendant can be kept isolated with the patient there will be less liability of carrying the infection through the house. In a majority of cases in families, however, the mother is obliged to care for the patient and to attend also to her household duties. In the last case, let her keep near the door of the room a cotton wrapper which can be put on over her dress whenever she enters the room. She had better tie a towel over her hair. In the room a china-stone basin should be kept, containing a gallon of water, in which there is a teaspoonful of the acid bichloride. Every time the attendant touches the patient let her wash her hands in this mixture, using no soap. She should remove her finger rings or they will be blackened. The patient should not be handled except when absolutely necessary, to avoid needless exposure to infection; it is also injurious to a child ill with diphtheria to lift it up. The nurse's covering wrapper should be soaked in the tub as often as possible. Some ignorant persons give as an excuse for a lack of care in handling patients having contagious diseases like diphtheria, that they are not afraid of the infection. Fear has nothing to do with the matter.

Food is to be taken to the door of the sick-room by some one other than the attendant. The tray should not be carried into the room. After the meal, take to the door a pan containing water, and let the attendant set the dishes, knives and forks, and the food handled by the child, under the water without touching the rim or sides of the dish-pan. Then any one may carry the pan to the kitchen, where it is to be set upon the stove, and the water holding the dishes and the rejected food is boiled for an hour. After that process the contents of the pan are safe, and they may be handled for washing. Cloths used in wiping the mouth of the patient are to be wrapped in paper and burned. Dejecta should be covered with fresh chlorinated lime, one part to two of water.

After the patient begins to convalesce the danger of infection grows greater. When the membrane has disappeared, and the child is able to run about the room, the attendant ceases commonly to use the throat-spray because the process is troublesome. In such cases the diphtheria bacillus remains in the patient's mouth for some time—from a few days to weeks. During the most of this time the child is as dangerous to others as it was while it was ill. In one case in my own experience, the bacillus remained present for eleven weeks from the date of diagnosis, and I then lost sight of the child. In the tenth week the bacillus present when in pure culture killed a guinea-pig in thirty-six hours. This is, of course, an exceptional occurrence; but the routine practice is to keep the patient isolated for three weeks after the membrane has disappeared, unless a bacteriological examination shows that the bacillus is absent. The bacillus remains after the use of antitoxin just as if antitoxin had not been used.

When a child is to be released from the sick-room, bathe it carefully with soaped warm water, washing out the hair and under the finger-nails carefully. Then wet a towel with the disinfectant (the acid bichloride of mercury,—a teaspoonful to a gallon of water) and go over the body with it; afterward rinse with ordinary water. Do not let the disinfectant enter the child's mouth or eyes. Next, without allowing the child to touch anything in the room, especially avoiding the door-knob, send it to another room and dress it in clothing that has not been near the sick-room. If, after this process, other children are infected, the explanation is that the child had been released too soon—before the bacillus had disappeared.

It commonly happens that a child has been going about the house for some days before a physician has been called in. In that event you have the house to disinfect. You must then wet with bichloride everything the child has touched, and boil all eating utensils.

As to the disinfection of the room and its contents: the irritation of diphtheria causes a large quantity of saliva to flow from the patient's mouth; this infected saliva runs down upon the pillows and soaks into them. It may also soak into the mattress. If a town has a steam disinfecting plant, there is no trouble in dealing with bedding and carpets after diphtheria and other contagious diseases; such a plant, however, costs at the least $6000. It is safer, in the absence of steam disinfection, to destroy pillows by fire; but if these are opened and the filling put into tubs or barrels containing two teaspoonfuls of the acid bichloride of mercury to each gallon of water and soaked for about two days they will be safe. The ticking in this case should be boiled in a wash-boiler, and the filling is to be rinsed before drying. The mattress is less liable to infection but it may be infected. If a piece of oil-cloth or rubber sheeting is spread beneath the bed-clothes under the patient and the mattress is kept well covered during the course of the disease, the filling of the tick will most probably be not infected. The loss of a good feather or hair mattress is considerable in the house of a poor man, and these often may be saved. To disinfect the surface of a mattress place it on chairs in a small room or in a closet and pour upon a cloth under it 500 cc. of formalin for each 1000 cubic feet of air-space in the room or closet—multiply the length by the height by the width of the room or closet to get the cubic feet of air-space. Leave the room or closet shut tightly for twenty-four hours. The Trenner-Lee formaldehyde disinfector is a good apparatus for disinfecting. The smaller size costs twenty-five dollars.

If anything is to be sent out of a room to be burned, spread a piece of old carpet, bagging, or similar useless cloth outside the room door, set on this the articles to be destroyed, wrap them carefully in the fabric, tying all with cords; then take the bundle outside the town in a covered wagon, pour kerosene oil on the package without opening it, and set it afire. Afterward wash the wagon with the acid bichloride.

Wet the furniture and floors of the room with the acid bichloride. Do not merely sprinkle the solution about, flood everything with it, because the germ is killed only by direct contact; and remember that a diphtheria bacillus magnified 800 times is not larger than the eye of a needle. The bichloride will spoil gilt picture-frames, therefore use a 10 per centum solution of pure carbolic acid on these and all other metallic surfaces. Coins should be boiled, and paper money should be dipped in the 10 per centum carbolic acid solution and dried at a stove. Money is frequently found in smallpox rooms under the patient's pillow.

Formalin is the best disinfectant for wall-paper unless the child has spat upon it—then use the bichloride. Sometimes the bichloride will not injure the wall-paper, but if there are gilt figures upon it these will be blackened. Sulphur fumes are no better than formalin—not so good, and they injure and blacken tinted and gilded wall-paper, silks, satins, and other fabrics. If you determine to have the room repapered, wet it with bichloride before you bring in the workmen.

It is difficult to disinfect a carpet except by steam, and on this account the carpet should be removed from the room before the patient is brought into it. If it has been kept in the room, wet it thoroughly with the bichloride, when you are disinfecting, if you can not have it disinfected by hot steam. The wetting commonly spoils the carpet, consequently it may be necessary to bum it.

Keep cats, dogs, and especially kittens, out of a diphtheria room. Kittens will take the disease easily, and cats and dogs will carry about the contagion. If a valuable dog should get into the room, disinfect its hair thoroughly with the acid bichloride and then rinse the hair. Be careful to disinfect its feet.

While using the bichloride do not forget the window-panes, the door-knobs, and that part of the chair-legs which touches the floor. After you have used the bichloride expose the room to the gas from formalin. Hang up sheets wet with 500 c.c. of formalin for each 1000 cubic feet of air-space, and close all keyholes and cracks; then leave the room shut for twenty-four hours.

As to the use of antitoxin as a preventive and cure for diphtheria, too much praise cannot be given to that wonderful discovery. Reliable diphtheria antitoxin, used in proper quantity and early enough, is almost an absolute cure. Where it fails it has been used too late or not in the proper dose. In any case its only evil effect may be an attack of nettle-rash or hives. The few deaths that have occurred in its use were caused by an ignorant use of the syringe. If you find a physician opposed to the use of antitoxin this simply means that he is a quack. One serious disadvantage in the use of antitoxin is that it leaves the dangerous bacillus in the throat of the patient about as long as an unaided convalescence would leave it. The membrane often will disappear in twenty-four hours where antitoxin has been used, and the child will be playing about the floor. Then the mother will say the child never had diphtheria; she will not disinfect, and she will let the child run about the house.

The free book system that prevails in some schools is a prolific source of infection. Books are infected at home or by children from infected houses, and mixed with other books in the school. The diphtheria bacillus will cling to a book for at least a year. If books are given to the children, give them outright; do not let the books be mixed in the schoolroom.

Drinking-cups used in common are another source of infection. Let each child have its own tin cup. The clothes-rack in a school also spreads infection. Room enough should be given to each hook to keep the hat and coat of one child from touching those of another, and a wooden partition standing out from the wall about eight inches should separate hook from hook. The janitor should wash the clothes-racks with the acid bichloride solution every time he sweeps.

Suppose a child having diphtheria is found in school, or one is discovered as coming from a house where he was in contact with diphtheria. The discovery is made commonly after the child has been spreading infection for some days. Do not frighten the youngster, but find out from him what parts of the school-building he has been visiting. Then send him and the other children home. Rooms in which the child has not been are not infected, and only that which he has touched is infected in any case. Wet everything in the building and outhouses with which he possibly could have come in contact with the acid bichloride. Burn his books and papers, or, if this action may cause difficulty with parents, let him take his books home and inform the health officer of that fact. When he returns to school be sure of the history of his books. Use formalin or sulphur in the infected rooms, and classes may be begun again the next day. If within the week any child shows signs of sore throat send it home immediately.

Sulphur must be burned when used as a disinfectant, and to be effectual four pounds should be burned for every 1000 cubic feet of air-space in the room. A teaspoonful of sulphur when burned will fill a house with choking, dangerous fumes, but two pounds of sulphur burned in an ordinary bedroom will have no effect whatever on the diphtheria bacillus and very little on any other disease. Sprinkling disinfectants about a house, and setting saucers containing disinfectants in rooms is nonsense—the quantity must be sufficient and be in actual contact with the contagion. A deodorant does not disinfect because it removes a stench.

To burn sulphur set a coal-hod or an old tin pan on two bricks in the middle of the room, but see that there are no holes in the bottom of the hod or pan through which burning sulphur could drip to the floor. For a like reason see that the pan is not too narrow nor too shallow. It is safer to set the bricks in a tub filled with water up to the top of the bricks. Use powdered sulphur in preference to the cakes sold by the druggists, and fire this sulphur with a red coal. The room should be moist with steam when the sulphur is set afire so that the fumes will act effectually. Leave it shut tightly for twenty-four hours.

In the Northern States diphtheria is most prevalent in October, November, and December; scarlet fever is an April disease, but it may occur at any time. It is easier to spread the infection of scarlet fever and measles than that of diphtheria, but it is not so difficult to disinfect after scarlet fever and measles as after diphtheria. The contagion of scarlet fever does not resist the fumes of sulphur or formalin. Disinfect a room after scarlet fever as for diphtheria but be sure to use also either sulphur or formalin because the contagion can float about a room. Eruptive contagious diseases like scarlet fever, smallpox, and measles so affect the skin that during convalescence the cuticle scales off. In severe cases of smallpox and scarlet fever the entire outer skin of the hand may peel off like a glove. The contagion is always found in the scaling skin. As the patient grows stronger the scales become finer, until at last they lie as mere mealy dust in the hollows of the elbows or other parts of the body. Down to the very last these scales are infectious, and they will retain the infection for months, probably for a year or more. The scales float in the air of a sick-room, fall on the clothing of visitors, are carried away by the shoes of those that leave the room. The scaling may continue for three weeks—it commonly does. These three diseases are infectious before the scaling begins, sometimes before the rash is well out. A very light attack of any of these diseases in one child may infect another fatally. Insist upon keeping a scarlet fever or measles patient out of school until all scaling has ceased.

Chickenpox is almost a harmless disease, but it is more infectious than even measles. Be cautious with it because nearly every epidemic of smallpox begins through some one mistaking smallpox for chickenpox, although there is little or no similarity between the diseases.

A child with tuberculosis of the lungs or a child infected with acute syphilis should not be permitted to go to school under any circumstance.

In the chapter on The Priest in Infectious Diseases will be found an account of the necessity of vaccination as a precaution against smallpox.

Tinea Favosa, or favus, is a contagious and a very stubborn disease of the skin, caused by the fungus Achorion Schoenleinii. It produces yellowish crusts about the hairs of the scalp and other parts of the body, and it destroys the hair. It attacks also the finger-nails and the skin that is without hair. In the later stages of the disease there is a foul odour. It is one of the most difficult of the scalp-diseases to cure; months and sometimes years are required to get rid of it.

A child with tinea should be kept away from school; and his desk and what he touches should be washed with the bichloride of mercury solution. Burn his books and papers.

Ringworm is a kind of tinea, and it is caused by various mould fungi. Tinea Tonsurans is ringworm of the scalp; Tinea Circinata is ringworm of the body; Barber's Itch is another form; there is also a ringworm of the finger-nails; and Pityriasis Versicolor is still another form. All are contagious, and some are difficult to cure because the parasite gets down between the skin and the hair-follicles and an antiseptic can not reach it. Children affected with these diseases should be kept away from school until they have been cured.

The presence of lice and of the Acarus Scabiei can bring about acute and severe skin eruptions. The Acarus Scabiei causes itch, but fortunately it is rare in America. These parasites go from person to person, hence a child having either should be kept from school until he is clean. A thorough washing will remove lice if they have not yet inflamed the skin, but itch requires a more vigorous treatment. The desks of such patients should be disinfected and their clothing should be baked. They will probably be reinfected at home if the treatment is not applied to other members of the family.

Contagious Impetigo, or porrigo, as it was formerly called, is a skin disease common among children, and it may affect adults. It appears to be of parasitic origin, but the specific organism that causes it has not been isolated. The lesions in this disease are commonly discrete—separate one from another—but they may be crowded together. They are vesico-pustular and they are sunken at the top in the typical form. If they are not broken by scratching, they dry into a yellowish crust. The disease affects only the skin, but as it is contagious a child affected with it should be kept from school until cured. The desk and articles used by the child should be disinfected, and his books are to be burned.

Whooping-cough is very infectious, and, contrary to the popular opinion, it is frequently a fatal disease. There is a period of incubation for from seven to ten days, then a catarrhal stage follows in which the child has the symptoms of an ordinary "cold." In about another week the dry cough becomes paroxysmal with the characteristic "whoop" when the air is drawn in after the fit of coughing. When there is an epidemic of whooping-cough, children with "colds" should be sent home from school. The objects used by a child that has whooping-cough should be disinfected, and its books and papers are to be burnt.

Mumps can be a serious and a very painful disease and it is infectious to a marked degree. The specific organism is not known. Boys are more liable to this disease than girls are, and recurrence is rare. After a period of incubation, which lasts from two to three weeks, there is fever, pain under one ear, and the parotid gland swells. The disease is commonly mild, but it may affect a child seriously. The patient is to be quarantined, what it has touched should be disinfected, and its books are to be burnt.

There are a number of infectious eye diseases that occur among school-children. Acute Contagious Conjunctivitis, or "pink eye," is one of the most important. One form of acute Contagious Conjunctivitis is caused by the Koch-Weeks Bacillus; it is "pink eye," properly so called, and it is very infectious. Objects handled by the patient can infect others and spread the disease. The attack is severe, but the prognosis for full recovery is good. The child should be strictly quarantined until all secretion from the eyes has ceased, and whatever he has touched is to be carefully disinfected.

Another form of Acute Infectious Conjunctivitis, less contagious than that caused by the Koch-Weeks bacillus, is brought about by the introduction into the eye of the bacteria that give rise to pneumonia. Commonly the pneumonia bacteria do not cause conjunctivitis unless the patient is susceptible in a special manner. As it is difficult to differentiate this second form from the first, the same precaution should be used.

Trachoma, called also granular conjunctivitis, Egyptian ophthalmia, and military ophthalmia, is a very serious inflammatory disease of the external eye which has of late years become prevalent in American cities, whither it has been brought by immigrants from eastern and southeastern Europe. Persons that have this disease on landing in the United States are deported, but despite this precaution it has crept in and is now endemic. It is contagious, and when well established it is extremely difficult to cure. If untreated it lasts for years and it may destroy the cornea and consequently the sight. A trachomatous child should be kept from school until it has been cured, and that cure will take a very long time.

The Gonococcus can be carried into the eye by handling objects like soap, towels, wash-basins, which have been used by persons afflicted with gonorrhoea. The infection of the eye is very severe and dangerous, and the usual quarantine is to be observed. The ophthalmia of the new-born is gonorrhoeal.

The Diphtheria Bacillus also may get into the eye, and set up a primary infection there. A membranous conjunctivitis, too, is at times induced by pus organisms. Xerosis Epithelialis, tuberculosis, leprosy, and syphilis may affect the eye primarily, and additional forms of eye-diseases are found that are infectious. The general rule, then, is that children with any inflammation of the eyes are to be kept out of school until a physician pronounces them harmless.

AUSTIN ÓMALLEY.

SCHOOL HYGIENE

Priests have to put up buildings for parochial schools, colleges, seminaries, orphan asylums, convents, and the like, but in such work sanitation is commonly given only a passing thought in connection with sewer-traps and these are left to the wisdom of a plumber. The physical welfare of youth is almost as important as its mental training, and there are many factors beside sewer-traps involved in the effort to sustain it.

If there is freedom of choice as regards the site of a schoolhouse or similar building, the top of a small elevation is to be selected. Such a position affords the best natural drainage, removes dampness, avoids inundations, gives full sunlight and the purer air. The top of a high hill may be too exposed to the wind.

Next to the top of a knoll, the southerly slope of a hill is to be chosen. The building should not be overshadowed by a hill, especially on the western side. Trees are not to be planted close to a building in which children live, and ivy and similar plants should not be permitted to cover the walls.

If a building is set in a hollow it will be surrounded with chill air and mists in the cold seasons, even if a costly drainage system keeps the cellar and basement dry.

A gravelly or sandy soil beneath a building is the best, provided this soil is not already saturated with organic matter, or is not close above a dense layer of clay or rock. Clay, marl, peat, and made soils should be avoided if possible, because they are full of organic matter; they are cold, and they infect the ground air. Rock does not make a good building site—its seams carry water.

The subsoil should be drained four or six feet below the cellar floor, and this floor is to be laid in concrete and cement. At the level of the ground there should be a course of hollow vitrified brick to exclude dampness and

173

to give ventilation.

Limestone walls conduct more heat in and out than an equal thickness of glass, bricks, plastering, and wainscoting. The porosity of the building material determines the interchange of the air through the walls, and it affects the temperature of the rooms. If there is water in the pores of the walls heat is conducted rapidly, but air is not permitted to pass. Brick as a building material has many disadvantages, but on the whole it is best for schools, and it resists fire better than most stones. The harder the brick the better it is—vitrified brick is the best. Hard-pressed brick of a light colour makes an excellent outer wall-surface.

It is very doubtful that sewer gas escaping into a house will directly carry the micro-organisms of diseases like typhoid and diphtheria, but such gas is poisonous, depressant, and it renders the inmates of a house liable to disease; lessens their power of resistance. The typhoid bacillus and other bacteria can, of course, be carried into a cellar by the seeping in of drainage water. Infants kept in the upper story of a house in hot weather are more liable to intestinal diseases than are those that live on the lower floors, but here the weakening agent is heat. Tuberculosis, scrofula, rheumatism, neuralgias, bronchial, and kidney affections are made worse in damp houses.

The chief defects in plumbing and drainage are the following: (1) Earthen pipe drains become broken or their joints leak, and they saturate the ground under a house with sewage. (2) Tree roots break and clog drain pipes. (3) The pipes sometimes have not fall enough. (4) Drains without running traps admit sewer gas. (5) Rats burrow along a drain pipe from the sewer into the house and admit sewer gas. (6) When the soil pipe from a water-closet is exposed in cold weather it may freeze up or be clogged by urinary deposits. (7) Rats gnaw through lead pipes and joints. (8) Two or more closets or sinks with unventilated traps on the same pipe will siphon back sewage. (9) Overflow pipes sometimes have no traps and they let in gas. (10) Ash pits near a house carry moisture to walls, (11) Cesspools leak through the soil.

In planning a school-building the classrooms and the study-halls are the first things to be considered. The classrooms should be oblong, with the aisles running lengthwise. Each child should have at the least 15 square feet of floor space and 200 cubic feet of air space. A room 30 by 25 feet with a ceiling 13 feet from the floor will serve for 48 pupils and no more. This is the best size for a room when blackboards and maps are used in teaching, because a larger room sets the children in the back seats too far away to see without eye-strain.

Dormitories should have at the least 300 cubic feet of air space for each child, and great care is to be taken in the ventilation. Children about 10 years of age require 11 hours of sleep; under 13 years, 10-1/2 hours; under

15 years, 10 hours; under 17 years, 9-1/2 hours; under 19 years, 9 hours. Do not make children get out of bed before seven o'clock in the morning; do not let them study before breakfast, and do not force them to work after half-past eight or nine o'clock at night until they are at the least 17 years of age. The hours for work should be:

Ages--From Hours of work a week
5 to 6 6
6 to 7 9
7 to 8 12
8 to 10 15
10 to 12 20
12 to 14 25
14 to 15 30
15 to 16 35
16 to 17 40
17 to 18 45
18 to 19 50

Work given for punishment must be included in these hours. No one, even an adult, should study for more than two hours at a time without an intermission for a few minutes. In a boarding-school no one under any pretext, even on rainy days, should be permitted to study during recreation hours, and the deprivation of recreation to make up lessons is a relic of barbarism. If a teacher can not get class work done except by shutting up children during recreation hours, remove the teacher or expel the pupil.

The amount of glazed window surface admitting light to a classroom or study-hall should be from one-sixth to one-fourth the floor space of the room, and this must be increased if the light is obstructed by neighbouring houses or trees. The light is to be admitted on the left side of the pupils,—all other windows should be counted as ventilators only. Windows facing the children or the teacher are to be avoided. In rooms fourteen feet high a desk twenty-four feet from a window is insufficiently lighted. The larger the panes of glass the better, and the external appearance of windows is to be sacrificed to good lighting. If screens are used to protect the glass from stone-throwing, allowance is to be made for the light the screens cut off.

If a room can not have enough light from the left side alone, put the additional windows on the right so that their lower sills will be eight feet from the floor; and be careful in this case that the light from the right is not brighter than that from the left.

Windows should have as little space as possible between them to avoid alternate bands of shadow and light. Set them up as near the ceiling as possible, since the higher they are the better the illumination; and they should not be arched at the top. The lower window sills may be about four feet from the floor. When window shades are used to cut off direct sunlight, they should be somewhat darker in colour than the walls.

If artificial light is used in boarding-schools in the study-halls, the best light is one that is as near in colour as possible to the white light of the sun, and ample, but not glaring. It should be steady, and it should not give out great heat nor injurious products of combustion. Hence the electric light is the best; after that, gas through Welsbach or Siemens burners. Well refined kerosene oil gives a good light, but it is always dangerous. Acetylene gas is now used in a safe apparatus, and it also is an excellent light.

No colour that absorbs light should be used on the walls. Pale greenish gray, nearly white, is the most satisfactory colour. There should be no wall paper, curtains, or hangings of any kind in a school or college building. The wall decorations should be as plain as possible, with no roughened places to catch dust.

Stairways are to be well lighted; they should be at the least five feet wide, and have landings half-way between each story. Diagonal or spiral stairways are dangerous. Steps with six-inch risers and eleven-inch treads are the easiest for children, but six-and-a-half-inch risers may be used in high schools and colleges.

Carbonic acid in the air of a classroom is an index of impurity. External air has about three parts of carbonic acid in 10,000 parts of air, and above seven parts in 10,000 is injurious. Each person exhales about fourteen cubic feet of carbonic acid gas in an hour. There is no easy method of determining the quantity of carbonic acid gas present in a room, and we must therefore arrange the ventilation so that about 3000 cubic feet of fresh air an hour will be supplied to each person in the house.

Beside carbonic acid there are other impurities in house air, as dust, micro-organisms of disease, exhalations from bodies, sewer gas, and the like, which accumulate and do injury when the ventilation is defective.

If every person in a house has 1000 cubic feet of air space, natural ventilation will suffice ordinarily, but artificial ventilation is needed in schoolrooms and dormitories. The subject of ventilation can not be satisfactorily discussed in a short article, and those that are interested in school building should leave the matter to a competent architect, or study books and articles like J. S. Billings' Ventilation and Heating, Pettenkofer's Ueber Luft in den Schulen, and Kober's article on House Sanitation in the Reference Handbook of the Medical Sciences.

The proper heating of a schoolroom is a matter so generally understood that there is no need for special remark here, except this, that provision for proper humidity in the heated air is commonly neglected.

Cheap water-closets do not save money—they get out of order too easily. The pan, valve, and plunger hoppers are not to be tolerated. The only kind to use are short-hopper closets with a trap that opens into the soil-pipe above the floor. These may have valve-lifters attached to the seats, because children forget to flush the hoppers. The ventilation of the water-closets should be separate from that of the main building. In country places where vaults are used, there should be a supply of dry loam kept, and enough of this to cover the fresh contents should be thrown into the vaults every evening.

Children are seemingly always thirsty, and they should be allowed to have all the drinking water they want if the source is free from typhoid germs and infection by organic matter. Common cups are an abomination, and a prolific cause of contagious diseases. Each child should have its own cup.

The rules for desks and seats for children are these:

1. The height of the seat should be about two-sevenths of that of the body.

2. The width of the seat should be about one-fifth of the length of the body, or three-fourths the length of the thigh. Do not keep unfortunate little children's feet dangling all through their school years to save a few pennies on school furniture.

3. The seat should slope downward a little toward the back, be slightly concave, and have rounded edges in front.

4. There must be a back-rest.

5. The child, when sitting erect, should be able to place both forearms on the desk without raising or lowering the shoulders. This is a very important rule.

6. The seat must be correctly placed as regards the distance of its front edge from the corresponding edge of the desk.

7. The desk slope should be 15 degrees.

Badly constructed desks cause eye-strain and marked distortions of the spine. Desks should be adjustable in height, especially for growing children. School-children grow most rapidly between the ages of twelve and sixteen years—nearly two inches a year—and the desks and seats should be adjusted twice a year at the least. If a child is moved to another desk an adjustment is to be made at once.

To counteract the bad effect of long sitting, even at properly adjusted desks, children should be frequently sent to blackboards, and at regular intervals a few minutes are to be given to "setting up" exercises.

Great attention should be paid to the eyesight of children. Those that complain of headache should have their eyes examined. The lines in school books should be not more than four inches in length, and they are to be printed in clear, well-leaded type. Slates are dirty and unsanitary: let the children write on paper that has a dull finish.

Teachers should prevent lounging positions at desks, especially stooping. They are not, however, to try to make children under fifteen years of age sit still. The youngsters can not remain immovable, and the effort to make them do so is irritating to no purpose.

Nervous children need outdoor exercise more than anything else. When nervousness takes the form of religious scrupulosity in school-children and novices do not immediately apply a moral theology to them—call in a physician that has common-sense, because there is a nervous scrupulosity which is much more frequently met with than the purely spiritual form. Aridity in prayer, a loss of sensible devotion, and similar troubles have to do with advance in the spiritual life, but they more commonly have to do with the liver in persons that are not nearly so important spiritually as they fancy they are; and in these cases the cook is the particular devil at fault, if they have exercise enough.

One of the chief sanitary evils in our boarding-schools, convents, and similar institutions, is the stupid sameness in the food which may be otherwise unobjectionable. The meat, for example, may be good, but the college and seminary cook sends it into the refectory chilled and clammy, or hot and overdone. In any case it is everlastingly the same. Children can predict a dinner's ingredients a month in advance.

Give children meat twice a day; white flour in their bread, because it is digested better than whole flour; all the sugar they want at meals; milk rather than tea, and tea rather than coffee; but let it be tea, not a dose of tannic acid.

The physical education of girls is neglected. Their general education is effeminate rather than feminine. If a convent faculty grows bold and "modern" it hires a teacher of gymnastics, puts an "extra" on the bill of expense, and ten or twelve wealthy girls play at gymnastics if they are not too lazy. Even if the whole school is obliged to attend the club-swinging and posturing and the other nonsense, little good is done. Girls should be kept out of doors for their exercise, and fresh air is much cheaper than a gymnastic teacher. If school-girls were forced into the open air more, they would not have time for munching caramels over the erotic spasms of Araminta and Reginald in the popular novel, and there would be advantage in the change. The absence of daily, regular, and sufficient exercise renders girls listless, anaemic, sallow, foul-breathed, melancholy, stooped, irritable.

Do not permit boys under eighteen years of age to go into regular training for college track-teams. Their hearts are not strong enough for the strain.

Boys should not use tobacco in any form, but it is useless to try to make them believe this statement. Tobacco stunts a boy, causes dyspepsia, and renders his mind dull. The measurements made for years at Yale, Amherst, and other colleges, by physical directors, show remarkable reduction in the height and chest expansion in tobacco users as compared with boys that do not smoke. Cigarette smoking would not be different from other smoking if it did not so readily tend to excess. Cigarette smoke is inhaled more than the smoke from cigars and pipes, and thus more of the injurious ingredients of tobacco are absorbed.

If a boy will smoke let him use a good long-cut tobacco which has little or no Perique tobacco in it, in a "Remington," "Edison," or similar wooden pipe. These are pipes with stems of large calibre, and in the stem there is a roll of absorbent paper or pith which keeps the pipe clean. Cigars, no matter how costly they may be, are too strong for a boy and for most men. A poor cigar irritates the throat aside from the regular effect of the tobacco, especially if there is much nitre in the wrapper. Meerschaum pipes are dirty and too strong. The tongue is irritated by a pipe that has a small bore in the mouthpiece: use a mouthpiece that has as large a bore as possible. Cigar smokers should, after cutting off the end of a cigar, blow the dust out of it from the lighting end to avoid inhaling this irritating dust.

AUSTIN ÓMALLEY.

MENTAL DISEASES

It is a well-recognised fact that persons suffering from many forms of beginning mental disease are likely to be affected by an exaggeration of religious sentiment. An unaccountable increase in piety is sometimes the first warning of approaching mental deterioration. It is not hard to understand why this should be, since religious feelings occupy so prominent a place in the minds of the majority of people, and the removal of proper control over mental operations of all kinds leads to an exaggeration, especially of those that have meant most for the individual before. Supposed religious vocations, especially when of sudden development, are sometimes no more than an index of disturbed mentality. Every confessor of lengthy experience has had some examples of this. This makes it important that clergymen should have a knowledge of at least the first principles on which the diagnosis of mental diseases is made. Superiors of religious communities, and especially those that have to decide as to the suitability of those applying for entrance to, or already in probation for, the religious life, need even more than others a definite knowledge of the beginning symptoms of the various mental diseases, and of the types of individuals that are most prone to suffer from them.

Besides, confessors and religious friends and advisers often gain the confidence of the mentally diseased much more fully than any one else. It is to them especially that the earliest symptoms of beginning mental disturbance are liable to be first manifested. After all, a pastor's and a confessor's duty is bound up with the welfare of his spiritual children in every sense; and it would be supremely serviceable to the patients themselves and to their friends, if these earliest symptoms could be recognised and properly appreciated, and due warning thus given of the approach of further mental deterioration.

The mental diseases that are of special interest in this respect are the so-called idiopathic insanities. Idiopathic is a word we medical men use to conceal our ignorance of the cause of disease. Idiopathic diseases are those that have come of themselves, that is, without ascertainable cause. As a matter of fact, the most important group of mental diseases develop without presenting any alteration of the brain substance, so far as can be detected by our present-day methods of examination. The initial symptoms of these diseases, then, are of great importance, and not readily recognisable unless looked for especially. There is no physical change to attract attention, and the change of disposition and mental condition is often insidious and only to be recognised by some one who is in the confidence of the patient. It is in these idiopathic insanities, then, that the careful observation of the clergyman is of special significance. Needless to say, powers of observation to be of service must be trained.

While there are no known changes in the brain tissues in these diseases, it seems not improbable that the development of our knowledge of brain anatomy, which is especially active at the present time, will very soon demonstrate the minute lesions that are the basis of these mental disturbances. It seems not unlikely that the underlying cause of so-called idiopathic insanity is usually some change within the brain cells. Hints of the truth of this conjecture are already at hand. Meantime the actual observation of this class of patients in asylums and institutions, private and public, and the collation of the observations of authorities in psychiatry from all over the world, have thrown a great deal of light on these forms of mental disease. We know much more of the initial symptoms and of incipient conditions that threaten the development of mental disequilibration than we did twenty-five years ago. With regard to prognosis especially, recent publications have added considerably to our knowledge, although it must be confessed that they have rendered our judgment of such cases much less hopeful.

The ordinary forms of mental diseases have sometimes been considered as passing incidents in the lives of patients suffering from such disorders. While it was generally understood that severe cases were apt to have recurrences, and that after persistence of mental symptoms for a certain length of time the outlook as regards eventual absolute cure is rather dubious, yet the general prognosis of such simple states as melancholia or simple mania was not considered to be distinctly unfavourable. Patients might very well recover their mental sensibility after even a severe attack, and never have a relapse.

It was something of an unpleasant surprise to the medical world, a few years ago, when one of the most distinguished authorities in Europe on the subject of mental diseases, Professor Kraepelin, of the University of Heidelberg, stated in his text-book of psychiatry, that among a thousand cases of acute mania he has observed only one in which the symptoms did not recur. Professor Berkley, of Johns Hopkins University, Baltimore, a conservative American authority, in discussing this subject of relapses after single occurrences of mania, is evidently of the opinion that Professor Kraepelin's opinion in the matter presents the inevitable conclusion that must be drawn from recent advances in the clinical knowledge of maniacal conditions. "Simple mania," he says, "is, according to the statistics now at hand, an exceedingly rare form of mental disease, and the physician should therefore be cautious in making a prognosis of final recovery. Relapses after a number of years, when stability is apparently assured, are frequent, as every one interested in mental medicine knows only too well."

The more experience the specialist in mental diseases has, the less liable he is to give an opinion that will assure friends of the patient that relapses may not occur after any form of disturbed mentality. While this is true in mania, it is almost more generally admitted with regard to melancholia. Most patients who have one attack of severe depression of spirits will surely have others if they are placed in circumstances that encourage the development of melancholic ideas. Any severe emotional strain will be followed by at least some symptoms of greater depression than would be expected from the normal person under the same conditions.

Professor Kraepelin has pointed out that in about one out of six cases the patients who came to him supposedly for the treatment of primary attacks of melancholia proved to be really suffering from a relapse of severe mental depression. The careful investigation of the history of these cases showed that they had suffered from previous attacks of depression, though sometimes these were so slight as not to have attracted any special attention from the medical attendant,—if indeed one had been called in the case—and at times even failed to occasion more than a passing remark on the part of friends with whom the patient was living.

The most frequent form of idiopathic insanity is melancholia. The disease is characterised by depression of spirits. Professor Berkley's definition, besides being scientifically exact, is popularly intelligible. According to him, "Melancholia is a simple, affective insanity in persons not necessarily burdened by neuropathic heredity, characterised by mental pain which is excessive, out of all adequate proportion to its cause, and accompanied by a more or less well-defined inhibition of the mental faculties." This latter part of the definition is extremely important. In extreme cases patients are able to accomplish no other mental acts beyond those which concern the supposed cause of their depression. Their lack of attention to other things is the measure of the mental disturbance. Their minds constantly revolve about one source of discouragement. They become absolutely introspective and their surroundings fail utterly, in pronounced cases, to produce any reaction in them. In milder cases this involves an increasing neglect of whatever occupation the patient may have, solely for the purpose of giving up time to the contemplation of the cause of his depression.

It is not easy always to recognise the limits between a depression of spirits that is not entirely abnormal and a corresponding state of mind that is manifestly due to insanity. When misfortunes occur, individuals will be mentally depressed. Sorrow has in it necessarily no element of mental alienation. It is only when it becomes excessive that observers realise that there is disturbance of the mental faculties, causing the undue persistence and the exaggeration of the grief.

For example, a mother loses an only son in the prime of manhood and at the height of his career. It will not be surprising if, for a considerable period, she is unable to take up once more the thread of life where it was so rudely interrupted. For weeks she may react very little to her surroundings and may prove to be so moody as to arouse suspicion of her mental condition. After a time, however, she begins to have some of her old interest in affairs around her. Her depression of spirits may not entirely disappear for long years, perhaps never; but her affective state does not go beyond a simple sorrow. On the other hand, under the same circumstances, a mother may give way to transports of grief that after a while settle down into a persistent state of dejection. Every thought, every word, every motive, has a sorrowful aspect to her. After a time she may begin to think and even to state that the misfortune of the loss of her son has come because of her own exceeding wickedness. She may consider it a punishment from on high and think that she has committed the unpardonable sin and absolutely refuse any consolation in the matter. This state of mind is distinctly abnormal, and if it persists for some time must lead to the patient's being kept under careful surveillance.

The immediate cause of the development of such a melancholic state is always some unfortunate event in the course of life. Worry and sorrow are important causative factors. Mostly, however, these causes are only capable of producing their serious effects upon the mental state of predisposed individuals, or at times when the health of the subject is decidedly below the normal. Emotional disturbances are not liable to have such serious effects, except when anaemia, or continued dyspepsia, or some serious nutritive drain upon the system, like frequently continued hemorrhages, persistent dysenteric conditions, or too prolonged lactation, have brought the system into a condition of lowered vital resistance. Unfortunately, in ordinary life these run-down physical conditions are prone to be associated with the worry and overwork that precede disaster.

The effect of grief as a cause of melancholia may best be realised from the fact that in something over one-half of all the cases of melancholia the death of a near relative, father or mother, or even more frequently husband or wife, or child, is found in the clinical history of the patient shortly before the development of the mental disturbance. Serious business troubles, however, loss of property, actual want of proper nourishment, failure to succeed in some project on which the mind has been set, and similar conditions, so common in our modern hurried life, are also capable of producing the mental depression that assumes an insane character in certain individuals.

For the development of melancholia a predisposition seems to be necessary. Most people can suffer the reverses of fortune, the accidents of life, and the griefs of loss of friends and relatives, without mental disequilibration. Certain predisposing factors are well known. Heredity, for instance, is extremely important. Melancholic conditions are frequently found in successive generations of the same family. While heredity is not as prominent a feature in melancholia as in other forms of insanity, the direct descent of a special form of melancholic mental disturbance from one generation to another is noted more frequently than in any other form of insanity.

Women are more often the subjects of melancholia than are men. This is especially true in the earlier and in the later periods of life. In the years between twenty and thirty-five the proportion of cases in each sex is more nearly equal. The two conditions, the establishment of the sexual functions, that is, the important systemic changes incident to puberty, and the obliteration of the sexual function at the menopause, with its consequent physical disturbances, are especially important in predisposing to the occurrence of melancholia in women. Their mental functions are less stable naturally, and are subject to greater physical strains and stresses. Childbirth and lactation are also important factors in the causation of the condition. Long-continued lactation—that is, beyond the physiological limit of about nine months—is especially a frequent cause. The development of the mental disturbance in this case is always preceded by a state of intense anaemia, in which the skin assumes a pasty paleness, and other physical signs give warning of the danger. Lactation is sometimes prolonged for no better reason than the hope to avoid pregnancy. Usually we may say this method fails of its purpose and pregnancy and lactation together work serious harm.

In young people particularly, homesickness is a not uncommon cause of melancholia. It is especially liable to produce the condition if young people at a distance from home are subjected to serious mental and physical strain at a time when the food provided for them is either insufficient or unsuitable, or when disturbances of their digestive systems make it impossible for them properly to assimilate it. A number of instructive examples of this condition have occurred in the last few years among our young soldiers in the Philippines. To the physical strain necessarily incident to campaigning, especially in young men unaccustomed to the life of the soldier, there was added the serious trial of the tropical climate and the unusual and not over-abundant or varied diet provided by the army rations.

Autointoxication is said to play a prominent rôle in the causation of melancholia. This supposes that there is a manufacture of poisonous materials within the system, whose transference to the nervous tissues causes functional disturbance of these delicate organs. Such poisons are especially liable to be manufactured when digestive disturbances have existed for long periods of time, or when chronic alcoholism is a feature of the case. The ordinary depressed condition so familiar in our dyspeptic friends and that develops so commonly as the result of indigestion, is an example of the depressing effect of toxic substances upon nervous tissues and mental states.

Melancholia does not develop as a rule without some warning of what may be looked for. Nutritive disturbances are nearly always prominent features in the case for some time before any mental peculiarities are noticed. Professor Berkley remarks that a feeling of woe and of uneasiness seems to be the way by which the brain expresses its sense of the lack of proper nourishment. Usually there has been distinct digestive disturbance for some months. There is apt to be loss of appetite. There may be some slight yellowness in the whites of the eyes. Commonly there has been an increasing disregard for the patient's usual habits, especially in the matter of exercise and friendly intercourse. There is a disposition to sit apart and brood by the hour, and a well-marked tendency to avoid friends and even members of the family, with an utter disinclination to meet strangers.

One of the marked features of the disease in women is a tendency to untidiness. Women lose all regard for their personal appearance and fail to arrange their clothes properly. Men who have been specially neat in their personal appearance take on slouchy, careless habits, allow their clothes to become soiled and dirty, and have evidently forgotten all their old customs in this matter.

The symptoms are not always continuous. There is often a rhythmic alteration of intensity of symptoms that corresponds more or less to the physiological rhythm of life. In ordinary circumstances human temperature is highest in the afternoon and vital processes are most active at this time. The lowest temperatures occur in the morning, especially in the early hours; and it is at this time that vital processes least active and the general condition is most depressed. It is not surprising, then, to find that melancholic patients are liable to suffer from deeper mental depression during the morning hours. In suicidal cases it is especially in the morning hours that patients need the closest surveillance.

In a certain number of cases of melancholia, instead of the quiet, often absolute immobility of the patients, there is a form of the disease characterised by the presence of incessant movement and an agitated state of countenance, that disclose their disturbed mental conditions. In melancholia, as a rule, sleep is very much disturbed, and at times patients do not sleep at all. In the agitated form of melancholia, the patient is often quiet only when under the influence of a sleeping-potion. Patients may tear their hair, disarrange their clothing, strike themselves, hit their heads against the wall, sigh and sob, and repeat some phrase that indicates their deep depression. They are apt to reiterate such expressions as "I am lost," "I am damned."

This is a much more serious form of melancholia than the quiet kind. The mental faculties are much more completely unbalanced, and the prognosis of the case is more unfavourable. There may be recovery within a very short time, and this recovery may be more or less complete. Usually, however, the condition becomes chronic and runs for many years. Such patients may sometimes be distracted sufficiently from their state of depression to smile and manifest pleasure in other ways. Usually, however, this diversion is only temporary and they recur to their darker moods until some new and specially striking notion distracts their thoughts once more.

With regard to melancholia the most important feature is the tendency to suicide. This is apt to be present in any case, however mild, and may assert itself unexpectedly at any moment. Where there is suspicion of the existence of melancholia, patients must be under constant surveillance; and, as a rule, they should be under the supervision of some one accustomed to the difficulties that such cases may present. Patients are often extremely ingenious in the methods by which they obtain the opportunities necessary for the commission of suicide. For instance, a man who has been calm in his depression and has shown no special suicidal tendencies may make his preparations apparently to shave and then use his razor with fatal success. In a recent case in New York City, a woman under the surveillance of a new, though trained nurse, asked the nurse to step from the room for a moment. When the nurse came back three minutes later, the woman was crushed to death on the sidewalk seven stories below. A male patient asks an attendant to step from the room for a moment for reasons of delicacy, and takes the opportunity to possess himself of some sharp instrument or of some poison. At times, during the night, patients rise up while attendants doze for a few minutes, and find the means to hang themselves without the production of the slightest noise.

These unfortunate suicides are happening every day. They are the saddest possible blow to a family. Only the most careful watchfulness will prevent their occurrence. Clergymen should add the weight of their authority to that of the medical attendant in insisting, when such patients are kept at home, that they shall be guarded every moment. As a rule melancholic patients should be treated in an institution. Their chances of ultimate complete recovery, and, more important still, of speedier recovery than at home are much better under the routine of institution life and the care of trained attendants.

Nearly three-fourths of the patients who suffer from melancholia will recover from a first attack under proper care. Subsequent attacks make the prognosis much more unfavourable. Not more than one-half will recover from a second attack, and, although melancholia is often spoken of as a mild form of intellectual disturbance, recurring attacks give a proportionately worse and worse outlook for the patient.

If the general condition of the patient, that is, the physical health, is very much run down when the mental disturbance commences, then the outlook is much better than if the mental disturbance should occur when the patient is enjoying ordinarily good health. Thin, anaemic patients, contrary to what might be expected, usually recover and often their recovery is permanent. The first favourable sign in the case is an improvement in physical health. This is very shortly followed by an almost corresponding improvement in the mental condition. When the patient has reached the normal physical condition, the mental disturbance has usually disappeared.

It is an extremely unfavourable sign, however, to have run-down patients gradually improve in physical health without commensurate improvement in their mental condition. This is nearly always a positive index that the mental disturbance will continue for a long while, may not be recovered from completely, or may degenerate into a condition of dementia with more or less complete loss of mental faculties.

The severe forms of melancholia are apt to be associated with delusions. Fear becomes a prominent factor, and the patient is afraid of every one who approaches, or concentrates his timidity with regard to certain persons or things. Delusions of persecution are not unusual, and this sometimes leads to homicidal tendencies. After enduring supposed persecution for as long as he considers it possible, the melancholic turns on his persecutors and inflicts bodily harm. The simplest actions, even efforts to benefit the patient by enforcement of the regulations of the physician, may be misconstrued into serious attempts at personal injury, for which the patient may execute summary vengeance. At times the hallucinations take on the character of the supposition that attempts to poison them are being made. The patient may conceal his supposed knowledge of these attempts until a favourable opportunity presents itself for revenging them. On the other hand, it is not an unusual thing to have melancholic patients commit homicide with the idea of putting friends out of a wicked world. The stories so common in the newspapers of husbands who kill wives and children, of mothers who murder their children, are often founded on some such delusion as this. A mother argues with herself, that her own unworthiness is to be visited on her children, and that they are to be still more unhappy than she is. Out of maternal solicitude, then, but in an acute excess of melancholia, she puts them out of existence and ends her own life at the same time.

When the melancholia is founded on supposed incurable ills in the body, patients are sometimes known to mutilate themselves, or to have recourse to alcohol, or some narcotic drug, in order to relieve them of their pain, which is mostly imaginary, and make life somewhat more livable during its continuance. Alcoholic excesses are especially common in cases of recurrent or periodical melancholia. Many of the cases of so-called periodical dipsomania are really due to recurring attacks of severe depression of spirits, in which men take to alcohol as some relief for their intense feelings of inward pain and discouragement.

One of the most characteristic symptoms of melancholia is the refusal to take food. Sometimes this refusal is the consequence of an expressed or concealed desire to commit suicide. In many cases the refusal of food is associated with the patient's melancholic delusions. If the patient is hypochondriac, food is not taken because the stomach is supposed not to be able to digest it, or because it would never pass through the system. At times the delusions are in the moral sphere and the patient is too wicked to eat, or must fast for a long period or perhaps for the rest of life, with the idea of doing penance. As a matter of fact the refusal to eat is associated with the lowered state of function all through the system, which is the basis of the melancholic condition. This causes loss of appetite and lowering of the digestive function with a certain amount of nausea even at the thought of food, so that it is scarcely any wonder that patients refuse to take food. Needless to say, they must be made to eat. This often requires the insertion of a stomach tube and forced feeding. And as it must be done regularly, it is accomplished much more easily at an institution than at home.

The other most common type of functional mental disease is mania. This is a form of insanity characterised by exaltation of spirits with a rapid flow of ideas and a distinct tendency to muscular agitation. It is almost exactly the opposite of melancholia in every symptom. Originally, of course, mania meant any form of madness. Then it became gradually limited to those forms of insanity which differed from melancholia. Now it has come to have a meaning as an acute attack of mental exaltation. It is necessary to remember this development of signification in reading the older literature on the subject of mental disturbance.

Professor Berkley calls attention to the fact that Shakespeare's statement, "Melancholy is the nurse of frenzy," may have been founded upon the observation that there are few cases of mental exaltation without a forerunning stage of depression. It is characteristic of the acuity of observation of the poet whose works have created so much discussion as to his early training, that this association of mental states, which became an accepted scientific truth only during the last century, should have been anticipated in a passing remark in the development of a dramatic character.

Melancholia precedes mania so constantly that it is not an unusual mistake in diagnosis to consider a patient melancholic when an outbreak of mania is really preparing.

Mania is sometimes said to break out suddenly. As a matter of fact there are always preliminary symptoms; though these are of such a general nature that they may have escaped observation. The patient's history generally shows that there has been loss of appetite and consequent loss in weight, commonly accompanied by constipation and headache with increasing inability to sleep. Usually these symptoms have been present at least for some weeks or a month or more. Then the patient brightens up. Instead of the brooding so common before, there is a tendency to talkativeness; the eye is bright; the expression lively; in the midst of his loquacity the patient becomes facetious and jocular. The backward before become enterprising. Undertakings are attempted that are evidently far beyond the power, pecuniary or mental, of the individual. Active employment is sought, and, where this fails, restless to and fro movement becomes the habit.

Friends notice this change in disposition, and also note a certain lack of connection in the ideas. There is apt to be a distinct change of disposition. A man who has been very loath to make friends before, now becomes easy in his manner toward strangers and takes many people into his confidence. In the severer forms motion becomes constant; the arms are thrown around; to and fro movement at least is kept up; the voice becomes loud and is constantly used. Patients can not be kept quiet, and, as a consequence of their constant movement, their temperature rises and loss of sleep makes them weaker and weaker until perhaps physical exhaustion ensues.

The causes of mania are not always so distinctly traceable as those of melancholia. Heredity is an important factor. This is, however, not so much a question of actual direct inheritance of mental disturbance from the preceding generation, as a family trait of mental weakness that can be traced through many generations. Direct inheritance of acquired peculiarities no scientific thinker now admits. Family peculiarities, however, are traceable through many generations. So striking a peculiarity as the possession of six fingers or six toes has been traced through a majority of the members of as many as five generations in a single family. And as has been said other family traits can be traced back in the same way.

It would not be entirely surprising, then, if mental peculiarities and a predisposition to mental disturbance should be also a matter of inheritance. It is well known now that the physical condition of the brain substance may have much to do with the intellectual functions. Injuries to certain parts of the brain may cause special changes even of personal disposition. In the famous crowbar case, in which an iron drill over four feet in length was driven through one side of the head, it was noted that the man, who had been somewhat morose before, was inclined to be more amiable afterwards, but also had a tendency to be bibulous in his habits.

German clinicians have recently pointed out that the existence of an excess of pressure on the frontal lobes of the brain, such as is produced by the presence of a tumour, may cause a tendency to make little jokes. This symptom is known as "Witzelsucht." It is considered of distinct significance and value in localising tumours of the brain. The question of the type of the witticisms and particularly a tendency to obscenity are noted as a special diagnostic aid in the recognition of the character of these tumours by at least three prominent German medical observers.

If modifications of the brain substance can produce changes of disposition and temperament, it is easy to understand how temperament and disposition may be a matter of inheritance. If we inherit a father's nose and a mother's eyes, the minutest conformations of brain substance may also be inherited. It is on these, to a certain extent at least, that the general outlines of the disposition depend. It would not be surprising to find, then, a disposition to mental unsteadiness as the result of the transmission of brain peculiarities. Here, as in everything else, there is question, not merely of parental influence, but of the inheritance of the family traits, some of which are skipped in certain generations.

When melancholia and mania are said to be due to heredity as one of the principal causes, the meaning intended is that in certain families the brain tissues are liable to be transmitted in somewhat impaired condition, and that through these brain tissues the mind will either not act properly, or under the stress of violent emotion, the loss of friends by death, or the loss of fortune, or serious disappointments in life, or a love affair, the already tottering mental condition will be overturned. In a word, it is not the direct transmission of insanity, but of a predisposition to the development of insanity under stresses and strains that is a matter of family inheritance. This is considered true now not only of mental but of all diseases. Not consumption, but the predisposition to it is inherited.

These considerations make clear how important this matter of heredity is. Physicians and students of anthropology are so much concerned about the increase of insanity as the result of the intermarriage of defectives that we are constantly reading in the newspapers of attempts at the legal regulations of marriage, so as to prevent further racial degeneration. Under present circumstances, any such legal regulation is probably impossible; but it seems perfectly clear that clerical influence should be brought to bear to discourage, as far as possible, intermarriage among those of even slightly disturbed mental heredity. Especially must any such idea as the possible beneficial influence of matrimony (for there are popular traditions to this effect) be unhesitatingly rejected and it must not be allowed to tempt those interested to look on such intermarriage with indifference.

Another and more serious question for the clergyman is that of the vocation in life of those who are weak mentally. By vocation is meant not only religious calling, but the occupation in life generally. Young people of unstable mentality and especially those of insane heredity should be advised against taking up such professions as that of actor or actress, or broker, or other life duties that entail excitement and mental strain. As far as possible they should be discouraged from taking up city life, and should be advised to live quietly in the country.

Mania is apt to follow certain severe infectious diseases in delicate individuals. Pneumonia, for instance, or typhoid fever or chorea, and sometimes consumption or rheumatism, may be followed by a period of maniacal excitement. Severe injury to the brain or the pressure due to the presence of a brain tumour, may also be a cause of mania. A certain number of good authorities in mental diseases have called attention to the fact that mania is a little more liable to occur in patients who are suffering from heart disease. By this is meant in persons who have some organic lesion of the valvular mechanism of the heart. This leads to disturbance of the circulation and interferes with cerebral nutrition, thus predisposing to functional brain disturbance.

While melancholia occurs very frequently in older people, mania is almost essentially a mental disease of the young. The vast majority of cases occur between the twelfth and thirty-fifth year. The subjects of the disease are usually those who possess what is called the sanguine temperament, that is, hopeful, enthusiastic people, easily excited and aroused, easily cast down. Mania is much more common in females than in males.

One of the important characteristics of mania is the super-excitation of the sexual faculty. In many individuals the first sign of their mental disequilibration noticed by friends is a tendency to sexual excess. This is true of women as well as of men, and the extent to which this may manifest itself is almost unlimited. At the beginning of the disease this symptom is often a source of serious misunderstanding, and may be the cause of family disruption. Usually, before there are any open insane manifestations, there are definite symptoms that would point to a pathological excitement in the sexual sphere.

One of the most striking characteristics of maniacal patients is the anaesthesia that often develops and is maintained in spite of the most serious injury. Because of this, maniacal patients should be guarded with quite as much care as those suffering from melancholia. I have seen a patient who, during an attack of acute mania, had put her hand over a lighted gas jet, holding it there until the tissues were completely charred. The burner was behind an iron grating, but she succeeded in reaching it. Neither from this dreadful burning itself, nor during the after dressings, did she complain of the slightest pain. Because of this anaesthetic condition and the consequent lack of complaint, maniacal patients often suffer from severe internal trouble without the medical attendant having any suspicion of its existence. There are few conditions that are more painful, for instance, than peritonitis, yet maniacal patients have been known to suffer and die from peritonitis, due to intestinal or gastric perforation, without a single complaint.

Unexpected death frequently occurs in mania because of the failure to recognise the existence of serious pathological conditions. Pneumonia may develop, for instance, without the slightest complaint on the part of the patient and go rapidly on to a fatal termination during the exhaustion incident to the constant movement, it being utterly impossible to confine the patient to bed. Meningitis may develop in the same way and proceed to a fatal issue without the patient's making any complaint or any sign that will call attention to its existence. In the meantime, the patient may be constantly in the wildest motion and so add to the exhausting effect of the organic disease.

The prognosis of acute mania is not unfavourable. Patients suffering from a first attack will recover completely in eight cases out of ten. Notwithstanding complete recovery, relapses are prone to occur whenever the patient undergoes a severe emotional strain. As a rule not nearly so much mental disturbance is required to produce a second attack as the first one, so that patients require great care. In a certain number of cases recovery is incomplete; persistent delusions remain, and there may even be some weakness of intelligence. Paranoia, as it is called, mild delusional insanity, may assert itself and then may persist for the rest of life. Notwithstanding this, patients may get along in life reasonably well, though their mental condition is decidedly below the normal.

In a certain number of cases, after the period of excitement disappears, a certain amount of dementia is noticed. This consists of a distinct lowering of the intelligence, though without the presence of any special delusion. This dementia progresses until finally there is a state of almost complete obliteration of the mental faculties. The prognosis as to life in cases of mania is very good. Very few patients die during an attack of acute mania. At times there is a development of tuberculosis that proves fatal, because of the restlessness of the individual. Pneumonia or typhoid fever may also prove fatal.

Besides mania or melancholia, there is a third form of functional mental disease, which is a combination of these two forms. It is usually spoken of as circular insanity. The patient has usually first an attack of melancholia, then an attack of mania, and then after an interval melancholia and mania once more. We have said that most cases of mania develop after a distinct stage of depression of spirits, so that successive attacks of mania take partly the character of circular insanity. This latter disease, however, is an index of a much more degenerated mental state of the individual than is either mania or melancholia alone. When it occurs, the prognosis as to future sanity for any lengthy interval is unfavourable. A series of attacks alternately of depression and excitement finally make it necessary to confine the patient to an institution.

As might be expected in this severer form of mental disturbance, heredity plays an especially important part in circular insanity. At least 70 per centum of the patients affected show a family history of insanity in some forms. In this disease direct inheritance of this particular form of mental disturbance is noticeably frequent. The patients who develop this form of insanity usually show marked signs of degeneration, even before any attack of absolute mental disturbance has occurred. Wounds of the head, alcoholism, and epilepsy are prominent factors in the production of circular insanity. This only means that the predisposition to mental disequilibration is so strong that but very little is required to disturb the intellectual equilibrium.

Fortunately, circular insanity is rare. In 40,000 cases of insanity in New York State, only 96 cases of this form were noted. Mild types of the disease are not, however, very rare. Many otherwise sane people have alternating periods of hopeful excitement and of discouraging depression, not momentary but enduring for weeks at a time, which are really due to the same functional disturbances that in people of less stable mentality produce absolute insanity. These cases are of special interest to the clergyman and to directors of consciences.

JAMES J. WALSH.

NEURASTHENIA

Neurasthenia, or nerve-weakness, "the vapours" of the old novelists and dramatists, is a very common malady, and it gives the clergyman trouble by the turmoil it causes in families, religious communities, in themselves, and elsewhere. Whether the condition is a distinct disease or not, and that question has been voluminously discussed, is not altogether an important matter, but that there is such a group of symptoms is unfortunately a weighty fact. It takes so many forms that it is bewildering, and therefore not readily reduced to unity.

The cerebral form often exists independently. There is such a thing as "brain fag," although many complainants may have very little material for the fag to work on. Often such a patient is robust, even an athlete, and his assertions meet with ridicule or abuse instead of treatment. If the patient is a woman she is not seldom called "hysterical." She is not hysterical. Hysteria, by the way, is as distinct a trouble as a broken leg, and far more serious, and not a synonym for perverseness, as the term is popularly used.

In the cerebral form, business, reading, study "go into one ear and out the other." The patient's memory fails him temporarily just when he may need it most, say, in a speech or sermon; a fly buzzing on a pane is a calamity and a source of profanity; a flat note in the choir-singing is ample reason for doubting the divine origin of the church, and every petty trouble that whisks its harmless tail across his floor makes him seek the table-top. I have known a whole convent of nuns, who were closely shut in, with bad ventilation and a worse cook, until all were more or less neurasthenic, almost disintegrated by the presence of a lamb sent in as a pet; not because of the bleating or any ordinary reason, but solely because of the hideous incongruity and indecency in the fact that the lamb was a male.

The cerebral neurasthenic makes rash, impetuous changes in his mode of life. He leaves a religious order because the coffee is weak, he resigns an important post in a bank because the president uses snuff, he abandons medicine for trade because the curate meddled in the treatment of two of his patients. He takes on anxiety, locks up the house six times over the same night; meals are eaten in awed silence by his trembling children; altogether he is an unmitigated nuisance.

He may get religious scruples. If he is a priest he takes an hour to an hour and a half to say a low mass, and most of that time is spent in searching the corporal for imaginary particles or in drying the dry chalice. He rereads his breviary until he is exhausted. Because moral theologians say that certain scruples are from the devil, he is convinced that the devil takes a particular interest in his case. The devil did probably take a special interest in his father's or grandfather's lack of scrupulosity, for his condition is commonly a result of alcoholism in an ancestor.

There are three chief types of neurasthenics: in one class is the person that appears robust, and is really so except in his nervous system, which lacks a governor. Such patients have little more than a troubled appearance to draw the attention of a chance observer to their condition.

A second class is made up of eloquent narrators of their troubles. They try all the physicians in turn, then the homoeopaths and osteopaths and similar quacks, and they add patent medicines prescribed by themselves. They are petulant, capricious, and despite their apparent energy they accomplish nothing.

The third class are silent, limp, clammy-handed; they are brought against their will to see the physician; they are sulky; bitter and unreasoning haters; inclined to melancholy. They may have a tendency even to suicide, but this is somewhat rare. Neurasthenics are not so liable to insanity as is popularly supposed, but such an outcome is possible in certain cases. If their vague fears go on into a more or less fixed delusion there is cause for anxiety lest insanity result, but care should be taken here to be sure the delusion is really irremovable.

Some neurasthenics are afraid to cross an open square or a wide street, others dread any closed apartment. Vertigo is common; so is insomnia. Insomnia is almost a constant symptom. The patient may have naps or he may have uninterrupted vigils. Sometimes there is a heavy but unrefreshing sleep. Sleepless patients are thrown into distracting rage by the barking of a neighbour's dog, the howling of cats, or the cackling of a successful hen, and they haunt the magistrates' courts in efforts to suppress such noises. They put cotton in their ears, wear heavy nightcaps, stop clocks, board up windows in search of sleep, which is not found.

These patients commonly have an enduring feeling of weight or constriction in the head, especially at the occiput,—a headache that is not actual pain. They also have vertigo, which is independent of any aural disease, and this is transient, showing itself on abrupt changes of position.

Another phase of neurasthenia is spinal. These cases have pain in the back and their legs give out. The back-pain is a diffuse ache, or it manifests itself on pressure at certain spots along the spine. There may be severe pain at the coccyx, especially in women. The walking may simulate paralytic forms if hysteria is mixed with the neurasthenia. Cardiac symptoms are often prominent, especially palpitation, but there is a nervous excitation of the heart rather than any definite lesion.

The gastro-intestinal symptoms are often important. Pain referred to the stomach and acidity are common, the tongue is coated, the faeces scybalous. Digestion is torpid. Sometimes there is nervous diarrhoea. A list of the belly symptoms described by some neurasthenics is interminable.

We often find a sexual form, which is the worst of all and the hardest to cure. It is commonly connected with masturbation. Such neurasthenics are shameless in the description of their nastiness. It is better to keep them from marriage unless they are cured, and they are not to be foisted off on any one as husband or wife to effect a cure. Allbutt says of them: "I fear that some of our 'criminal psychologists' are encouraging many sorts of prurient debauchees by dignifying the tales of their vice with the name of science, a course of conduct which is in the worst interests both of these persons themselves and of our own profession. It were a curious inquiry how it comes that sexual perversions are so 'scientific' a study, while the brutalities of the thieves' kitchen or the wiles of other pests of society lie in comparative neglect."

Physical, intellectual, or emotional strain can cause neurasthenia suddenly or gradually. Where it comes on without obvious cause there is commonly a bad family history of nervousness or alcoholism. Anaemia makes it worse; eye-strain, too, is a provoking factor. In some cases a renal congestion is the cause. In many cases a lack of restraint, bad education, uncontrolled passion, are a marked influence in fixing the neurasthenic habit. A sedulous parent nags at a neurasthenic child that is too weak for exertion until the child's susceptibility to correction is blunted. Instead of treatment and help the child receives cuffs and abuse, and hell-fire is held up before him until he deems all religious talk dust and ashes. Encouragement will sometimes do more good than all the threats in the via purgativa. Nagging never cured anything except a tendency toward virtue, and it always deepens neurasthenia. Be careful in the selection of a confessor for a neurasthenic child. Get one that does not believe in kicking a soul into paradise.

The treatment of neurasthenia is difficult. Traveling about in search of health is not advisable. The Weir Mitchell Rest Cure is very effective in many bad cases, but it is costly, and if not correctly applied it is useless. It is the only cure for some patients. Sea air helps a certain class of neurasthenics, but it makes others worse—it is bad for the dyspeptic neurasthenic. A chronic rhinitis, a refractive error of the eyes, a displacement of the uterus, a congested kidney, a floating kidney, a tight prepuce, and similar teasing disorders must be cured before the neurasthenia can be removed; often the neurasthenia disappears with this cure.

Traumatic neurasthenia is like simple neurasthenia in most details. It is called also nerve shock, spinal irritation, railway spine. There is always a causative shock or injury, which is followed at once or after an interval by the symptoms of neurasthenia. In acute traumatic neurasthenia there may be, in addition to the symptoms observed in simple neurasthenia, high fever, and such a fever has been observed to go as high as 113 degrees Fahrenheit.

AUSTIN ÓMALLEY.

HYSTERIA

The term Hysteria (uterus) has been handed down from the days when physicians thought there was a connection between womb-disorders and the set of nervous symptoms grouped under the title hysteria. It is now etymologically meaningless,—men also grow hysterical. Briquet found 11 male to 204 female hysterics, and later statistics increase the number of males.

The disease is not readily definable. The patient is usually a young emotional woman, oftenest between 15 and 20 years of age. She commonly has anaesthetic spots on her body, concentric limitations of the field of vision, and hystero-genetic zones, or tender points, which, when pressed, appear to inhibit the hysterical fit. The symptoms enumerated here are not, however, found in every case of hysteria, and it is difficult at times to diagnose the disease.

The various manifestations of hysteria are (1) apt to come and go suddenly. A severe paralysis that suddenly disappears for a time is hysterical; (2) even if they last for years they may be suddenly cured; (3) they are dominated more by mental and moral influences than are the symptoms of any other disease; (4) we find no organic lesion with which we can connect the symptoms.

The conditions that bring about hysteria are hysteria in a parent, or insanity, alcoholism, or some similar neurotic taint in an ancestor. There is no direct connection between hysteria and the disorders of the sexual organs.

Immediate causes are acute depressive emotions, shocks from danger, sudden grief, severe revulsions of feeling, as from disappointment in love; and, secondly, cumulative emotional disturbance, as from worry, poverty, ill treatment, unhappy marriage, or religious revivals. Certain diseased conditions, as anaemia, chronic intoxications, pelvic trouble, cause hysteria, or, more exactly, start it into activity where it is latent. It is also communicated by imitation and it may become epidemic.

After the great plague, the Black Death, in the fourteenth century, there were very remarkable epidemics of imitative hysteria in Germany and elsewhere. In 1374, at Aix-la-Chapelle, crowds of men and women danced together in the streets until they fell exhausted in a cataleptic state. These dances spread over Holland and Belgium and went to Cologne and Metz. It is said that in Metz there were 1100 of the dancers seen at the same time.

"Dancing Plague" broke out again in 1418 at Strasburg, in Belgium, and along the Lower Rhine.

"Viel hundert fingen zu Strassburg an
Zu tanzen und springen Frau und Mann,
Am offnen Markt, Gassen und Strassen;
Tag und Nacht ihrer viel nicht assen,
Bis ihn das Wüthen wieder gelag.
St. Vits Tanz ward genannt die Plag."

Beckmann (Historia des Fürstenthums Anhalt. Zerbst. 1710) tells of a similar outbreak in 1237, wherein nearly a hundred children were seized by the disease at Erfurt, and they went along the road to Arnstadt, dancing and jumping hysterically. A number of these children died of exhaustion. The same infection is often at work in the fury of a mob, the panic of a beaten army, and it probably was an element in the Children's Crusade.

The Tarantism so common in Italy from the fifteenth to the eighteenth century is another example of epidemic hysteria. The Bubonic Plague ravaged Italy sixteen times between 1119 and 1340, and smallpox was at work when the black death could find no fresh victims. As a consequence of economic disturbance and fear the people were generally neurasthenic, and a slight shock was enough at times to set whole villages into hysterical convulsions.

In 1787, at Hodden Bridge in Lancashire, England, a girl in a cotton mill threw a mouse upon another girl that had a great dread of this animal. The frightened girl was thrown into a hysterical convulsion which lasted for hours. The next day three girls that had watched her were in convulsions, the following day six more, and two days later fourteen more girls and a man were in fits. American white and negro camp-meetings result in similar outbreaks, and the French Convulsionnaires, who did outrageous things from 1731 to 1790, were also afflicted with imitative hysteria. The Cornish Jumpers, founded in 1760 by Harris Rowland and William Williams, and the American Barkers were also hysterical. The Barkers in the meetings would run about on all fours growling, "to show the degeneration of their human nature," and they would end in almost general fits of imitative hysteria.

There was an epidemic of hysteria in Tennessee, Kentucky, and a part of Virginia, which began in 1800 and lasted for a number of years. It started at revivals. The majority of the cases were in persons from 15 to 25 years of age, although it was observed in every age from 6 years to 60. The muscles affected were those of the neck, trunk, and arms. The contractions were so violent that the patients were thrown to the ground, and their motions there exactly resembled those of a live fish thrown out of the water upon the land.

There are numerous theories formulated to explain hysteria; some are ingenious, especially that of Janet, but none is convincing. Convulsions, tremors, paralyses of various forms and degrees are common in hysteria. In major hysteria the patient falls into a convulsion gently. There is checked breathing, up to apparent danger of suffocation. Then follows a furious convulsion, even with bloody froth at the mouth, but there is a trace of wilfulness or purpose in the movements. Next may come a stage of opisthotonos, where the body is bent back in a rigid arch till the patient rests on her heels and head only, and this is followed by relaxation and recurrence of the contortions. An ecstatic phase succeeds this, at times in the so-called crucifix position, with outbursts of various emotions, and a final regaining of a normal state. Any of these stages, however, may constitute the whole fit.

In minor hysteria there is commonly a sensation of a rising ball in the throat (the globus hystericus). There may be uncontrollable laughter or weeping. Muscular rigidity is frequently found. The patient, especially if she is a child, may mimic dogs and other animals. The snarling, biting, and barking of false hydrophobia are hysterical; these symptoms do not occur in real hydrophobia.

There are almost innumerable physical symptoms of the disease, which are chiefly of medical interest, but the mental phases are such as to involve questions of morality. The hysterical character is marked by an overmastering desire to be an object of general sympathy, admiration, or interest, rather than by a tendency to baser indulgence. The will is weak, the emotions explosive, the patient is impulsive and lacking in self-control. She is a "giggler," who goes from absurd laughter into floods of tears. The desire for sympathy and attention makes the patient exaggerate her symptoms or simulate diseases and conditions that do not exist in her case. Hysterics will swallow pins or stick them into their flesh to force attention. Sometimes the simulation of disease is not willed. If there are a number of hysterical girls in a hospital ward and one develops, say, a peculiar paralysis, within two or three hours every hysterical woman in the room will have the same paralysis,—not pretended, but real, although temporary. It must be remembered that the disease, with all its perversity, is as much a fact as pneumonia, and the element of sham is only one of its symptoms. Some authorities go so far as to hold that a woman who will not lie is not hysterical. They invent most extraordinary slanders against even their own immediate family, and it is never prudent to believe an accusation made by an hysterical patient, no matter how plausible the story.

Acquired hysteria in many cases may be cured, but the congenital condition is practically hopeless, yet the latter kind may be kept from violent outbreaks.

We can not prevent drunkards, epileptics, and lunatics from propagating their kind, and therefore we shall still have the hysteric with us. The child that has a bad ancestry and shows hysterical tendencies should be carefully reared. If it has an hysterical father or mother it should, if possible, be removed from this evil influence. Keep it from long hours of mechanical work that leaves opportunity for dreaming. Shut out novels and "art for art's sake," especially music. Give it a practical education. Teach it obedience, self-control, and truthfulness. Harden its will by exercise at things it does not like, and do not coddle it. Do not marry off an hysterical girl to cure her. Do not inflict her presence upon some unfortunate young man because he is a good citizen. Marriage will not cure hysteria,—the worst cases are married women, and they beget other hysterics in spreading succession. When the disease shows itself offer no sympathy,—do not try to put out a fire with oil. When a "good, pious girl" grows hysterical, the chief obstacles to her cure are untactful and sympathetic visits from friends, lay and clerical. A visit from the pastor, because of his importance, is always harmful, and if the bishop drives up in his carriage so that the neighbours may see him, all the physicians in the city can not help her. If you wish to keep an hysterical girl in her vapours, get her a physician that will grow excited over her, take the dear child out of school and weep above her couch, let the family and its friends assure the unfortunate attending physician in her presence that he is heartless, and she will stay hysterical to her soul's content.

If you wish to control the attack, or even remove the disease under certain conditions, call in an experienced physician, leave the treatment to him, and pay no attention to her. Do not make light of the disease, do not speak of it at all. There are attacks that may be cured by the razor-strop or a bucket of cold water, but these are exceptional. They are new cases or old professional offenders. Rough treatment is not so good as patient tact, but at times roughness is the only cure.

AUSTIN ÓMALLEY.

MENSTRUAL DISEASES

Menstruation is a periodic discharge of blood from the uterus and the Fallopian tubes. It occurs every twenty-eight or thirty days, and it lasts from puberty to the menopause, or the cessation of the menses,—about the forty-fifth year of age.

There is a connection between menstruation and the production of the human ovum. During the first stage of menstruation the mucous membrane lining the uterus swells to twice or thrice its normal thickness, and this growth is a preparation for the reception of the ovum, which, as a rule, is given off by one of the ovaries at this time and passes out into the uterus. Menstruation and ovulation ordinarily occur simultaneously, but they may be independent and take place at different times. If, during this stage, the ovum is impregnated, pregnancy begins, and menstruation ceases until some time after childbirth. In married women conception is more likely to be effected during the first stage of menstruation than during the interval of quiescence; the contrary is almost the exception. Impregnation, however, is likely to occur in the spring more than at other seasons, and this fact coincides with the advent of spring in various latitudes.

If the ovum is not impregnated, the material that made the uterine mucous membrane thick during the first week of menstruation degenerates and passes off, constituting the menstrual flow. This stage lasts about five days. A reparative period of about four days follows, and then a period of quiescence until the next menstruation commences.

Menstruation is first observed about the fourteenth year, but it may start earlier or later. In general, it comes on earlier in warm climates, and later in the extreme north. The menstruation, too, is likely to show sooner in the labouring classes than in girls who do not work.

Even in normal menstruation there is often a marked physiological excitation which affects the entire person. Very commonly a nervous disturbance and sensitiveness are observed, and in women that are not robust there may be mental depression and irritability. The temperature will rise a half degree, and drop to the normal height on the day preceding the flow.

There are derangements of menstruation which are symptoms of various diseases. Amenorrhoea is an absence of menstruation in conditions other than pregnancy or lactation. Absolute amenorrhoea is a complete absence of menstruation for several months; relative amenorrhoea is delayed, scant menstruation.

Amenorrhoea is common during convalescence from acute diseases; it is also a result of chronic diseases of the liver, stomach, intestines, kidneys, and especially of the lungs; it complicates anaemia, malaria, rheumatism, and other general pathological conditions. Fright, grief, great anxiety, mental shock cause amenorrhoea; so do homesickness and many forms of insanity.

There are also local causes of this condition: imperfect development of the uterus or the organs connected therewith, and inflammations of these organs or of the pelvic wall.

Opposed to amenorrhoea is menorrhagia, or an excessive menstrual flow. Metrorrhagia, or hemorrhage from the uterus at any time, is a term confounded with menorrhagia, which is an inordinate menstrual loss of uterine blood, but the distinction is not important. Menorrhagia and metrorrhagia commonly have an identical cause and they frequently coexist. They are found in chronic diseases of the heart, lungs, liver, and other organs; they are an outcome of prolonged lactation, and of local affections of the uterus and its appendages. Any condition also that deranges the blood may cause menorrhagia or metrorrhagia; so do malignant tumours of the uterus, uterine displacements, lacerations that occur in childbirth, and psychical influences, as fright, anxiety, and other strong emotions.

Dysmenorrhoea, difficult or obstructed menstruation, is a term used for menstruation accompanied by pain. This is a common menstrual derangement, and it may be neuralgic or inflammatory in origin, or it may be caused by obstruction to the menstrual flow. There is another variety of dysmenorrhoea, called membranous, in which the superficial layer of the uterine lining is cast off partly or wholly.

In the neuralgic form the uterus and its appendages are normal in appearance, but the pain recurs monthly, and it may have degrees from mere discomfort to agony. This form is characterised by reflex headache, sympathetic nausea or vomiting; and the pain may not be confined to the uterus and its appendages. The irritation often brings out latent hysterical phenomena, spinal irritation, and neurasthenia. Rheumatism and gout are predisposing causes, so are indolence, lack of physical exercise, light clothing in cold weather, forced school work and similar depressing agents.

In the neurotic variety of dysmenorrhoea pain often persists after the menstrual flow has set in, but in inflammatory dysmenorrhoea the flow relieves the pain or removes it. Marriage commonly removes the neurotic form of dismenorrhoea.

In obstructive dysmenorrhoea the menstrual fluid is retained by narrow or tortuous outlets, flexions of the uterus, and similar causes. The prognosis is good in all forms of dysmenorrhoea, but frequently long and skilful treatment is required to cure such conditions, especially the membranous form. Inflammatory, obstructive, and membranous dysmenorrhoea are commonly made worse by marriage.

At the end of the childbearing period menstruation gradually ceases. In temperate climates this menopause occurs about the forty-fifth year, but it may come earlier or considerably later. Work that keeps a woman in a heated atmosphere, as cooking, washing, and baking, disturbs menstruation and tends to advance the menopause. Workers in chemical factories, in badly ventilated rooms, or women that do heavy labour in the open air, are apt to age prematurely, and have an early menopause or "change of life." This premature climacteric is found also in women that bear many children in rapid succession.

At the menopause there may be various physical or mental disturbances which are probably due more to the somewhat abrupt advent of old age, at the cessation of the childbearing part of life, rather than to the menopause itself. It is a fact, however, that often profound disturbances coincide with the climacteric, and we know no sufficient cause for them if the menopause itself may not be deemed such.

There are numerous disorders of the nervous system in women which are dependent directly or indirectly upon a derangement of the pelvic organs. Distant parts of the body are affected pathologically through sympathetic irritation when the primary disease is in the pelvic organs, and direct treatment of the pelvic trouble alone cures these reflex conditions. The very common disorders of pregnancy, the marked physiological changes in women at the beginning of menstruation with puberty, and its cessation with the menopause, are among the first proofs of this assertion that occur. Menstruation may aggravate goitre, uterine fibroid tumours, skin diseases, and affections of the blood vessels. Disordered menstruation causes sleeplessness, melancholy, dementia, and mania, by affecting the brain; it may bring on local paralysis; start up latent epilepsy; excite reflex cough and difficulty in breathing; make the heart irritable; cause nausea, vomiting, dyspepsia, flatulence, diarrhoea, skin-inflammations, pain in the joints, and many other symptomatic phenomena.

Chorea ("St. Vitus's Dance") is caused by various irritatations, and dysmenorrhoea can be such a cause. If a person is disposed to hysteria by neurotic inheritance, idleness, sedentary habits, vicious practices, excessive development of the emotions, any affection of the uterus or its appendages will greatly aggravate the outbreaks. The same is true in neurasthenia; and uterine disorders can directly cause neurasthenia, a condition described in another chapter. Migraine is an extremely severe form of headache which arises from various excitations, and uterine disturbances are among the causes.

Insanity frequently appears in women at puberty, soon after marriage, during pregnancy or lactation, and at the menopause; at these periods disposed women are especially prone to outbreaks of insanity. Irritation and exhaustion from diseases of the pelvic organs are potent factors in bringing on insanity, although these conditions may coexist independently of each other. Symptoms should not be mistaken for causes, but pelvic diseases at least aggravate a tendency toward mental unbalance.

In an article like this it is not expedient to speak of treatment, but the conditions are described in outline so that the spiritual adviser may recognise the need of medical aid and suggest its employment. A woman suffering from pelvic disorders should be relieved from a labourious or responsible office until she has been cured of her disease, in her own interest and especially in the interest of those affected by her condition.

AUSTIN OMALLEY.

CHRONIC DISEASES

It is often of great practical importance to bear in mind that a number of affections, commonly not serious in themselves at the beginning, and sometimes giving very few external symptoms, may make the mental condition of the individual suffering from them utterly incapable of meeting grave responsibilities. This is especially true with regard to such positions as that occupied by the Superior of a religious community who may, during the course of an ailment that has a tendency to affect the mental condition, do things that involve the community financially, or make life so uncomfortable for their subjects as to cause them to abandon the religious life. Some of these ailments are very insidious and may develop utterly apart from all anticipation in persons that were previously healthy. The weight of responsibility itself may, by impairing the general health, bring on an aggravation of a previously mild chronic condition that will cause distinct mental deterioration, yet without the absolute production of such disturbance of intellection as will be readily recognised by those that are not brought intimately in contact with the individual.

Such cases are not uncommon in history. A distinguished specialist in mental diseases called attention, in the London Lancet not long ago, to the case of Nicias, the Greek general who was in charge of the Athenian expedition against Syracuse. Nicias undoubtedly had a genius for war and for politics when in normal health. Some of the mistakes committed by him, though, are of an order that indicate a lapse of mental control at certain times. Details given by a number of Greek historians point to the existence in Nicias of symptoms of chronic nephritis, which at periods of great responsibility became exacerbated with consequent interference with normal intellection. The same authority points to certain otherwise inexplicable political mistakes in the life of Napoleon III. as due to the existence in him of a low-grade nephritis, consequent upon the presence of stone in the kidney. After his abdication, during his life in England, he had to be operated upon for this condition, and the calculi found had manifestly been in existence for many years.

Even more important for the sake of the individual himself than for those he is in contact with is the recognition of his pathological condition. Nothing is more likely to cause kidney disease to grow rapidly worse than responsibilities heavier than the individual is accustomed to. When, then, there are symptoms of nephritis it is inadvisable for the patient to be made Superior, and if the symptoms develop after his appointment or election he should be relieved of his responsibilities, at least to a considerable degree. There are a number of cases on record in which failure to realise the necessity for this mode of action has been a cause of great unhappiness in religious communities, and not infrequently a shortening of a very precious life that might otherwise have been spared for long years of usefulness in some less demanding position. It is not impossible that paresis should develop in the Superior of a religious community. The disease is extremely rare among clergymen generally, and the statistics of asylums show that it is rarest of all among Catholic clergymen. Should it occur, however, it must constitute a quite sufficient reason either for a change of Superiors, or for the institution of such other safeguards as may, according to the special religious institute, be provided in order to prevent serious evil.

In the religious communities of women, particularly, it has seemed to us that the occurrence of Graves' disease (the affection is three times more frequent in women than in men) in a Superior should always be the signal for relieving her of the responsible duties of her position. This action is quite as necessary for the patient's own health as for the peace and happiness of the community. The disease may exist in a latent form and only develop strikingly after the assumption of the serious responsibilities of the position of Superior. When, however, the eyes are prominent, the pulse rapid, and the goitre, or swelling of the front of the throat, characteristic of the disease, is present, there are practically always mental symptoms that make it extremely inadvisable for her continuance in a position of serious responsibility. Professor Church of Chicago (Professor of Nervous and Mental Diseases and of Medical Jurisprudence, in the Northwestern University Medical School), in the last edition of his book on Nervous and Mental Diseases, [Footnote 5] has this to say with regard to the mental disturbances of Graves' disease:

[Footnote 5: Nervous and Mental Diseases. Church and Peterson, 4th edition. Saunders, Phila., Pa., 1903.]

"From the beginning, and often for a long period antecedent to the appearance of cardiac symptoms, the subjects of Graves' disease present a considerable mental erethism. There is an indefinable and tormenting agitation, marked by mental and motor restlessness and an imperative and impulsive tendency to be doing. Their emotions are too readily excited, and they are unusually impressionable and irritable, reacting in an exaggerated manner to all the incidents of daily life. In more pronounced cases they become voluble and manifest the greatest mobility of ideas, but have no persistent concentration of logical order. Their affections are likely to undergo modifications, and they become irascible, fault-finding, inconsiderate, ungrateful, and hard to live with. In some instances this disturbance of mentation carries them over the border into active mania, marked, perchance, by delusions of fear, due to the cardiac symptoms of sensations of heat. Insomnia is often added and the fitful sleep is disturbed by horrifying dreams that are likely to be projected into the waking moments and woven into delusions which are usually unsystematised, and constantly changing, furnishing the analogue of the motor restlessness. Hallucinations of sight and hearing are not uncommon.

"The mental perturbation only rarely reaches the degree of actual mania, and then is, perhaps, equally dependent upon numerous other causes acting in a neurotic individual. But a condition of abnormal mental stimulation is characteristic of the malady, and is as important an index as any of the cardinal triad." [Footnote 6]

[Footnote 6: Of physical symptoms, namely, the rapid heart, the prominent eyes, and the enlargement of the thyroid gland in the neck.]

Dr. Church considers, then, that the mental symptoms of the disease are as important a concomitant, and as little likely to be absent in any given case, as are any of the three or four well-known physical symptoms characteristic of the disease. Under these circumstances the necessity for the exercise of care in permitting such a patient to continue in the office of Superior must be manifest. It is a question not for religious authorities to decide but for physicians, and they are to be experts in mental diseases. There are many physicians who have had experience with cases in which Graves' disease has been a source of unfortunate conditions in religious life, owing to the failure to understand the relations of the physical affection to mental disturbances. At times unfortunate consequences follow that are irretrievable in the destruction of vocations and the impairment of the religious spirit in communities.

As a rule it may be said that the development of serious disease is almost sure to incapacitate a Superior from fulfilling the functions of office. This is true, however, not only for physical disease but for the so-called neuroses. These are maladies which have their basis in some disturbance of the physical constitution, though this is not always easy to find. We prefer to speak of them as neuroses rather than neurasthenia, because this latter name has somehow come to have an unwelcome sound and to carry with it the idea of imaginary rather than real ailments. A true neurasthenic, however, is supremely to be pitied.

It has often been noticed that such individuals, while perfectly capable of judging properly for others, are not able to form right judgments with regard to their own conditions. This principle, however, should not be taken as a rule, and it must not be forgotten that neurasthenics are often the subjects of compulsory ideas—so-called obsessions, in which they are not entirely responsible for actions performed. At such times they are prone to be irritated by very trivial faults, and what is worse, to exaggerate slight defects into serious infractions of rule or of obedience. With regard to such persons, therefore, constant care has to be exercised to control their statements by those of others and not to take them at their full value without due substantiation. In this matter the subject is quite as likely to suffer as the Superior, and information obtained from them should not be acted upon without consultation with others who know the details of the case.

As a rule neurasthenic individuals become, as is well known, worse as far as the mental condition is concerned when they are asked to assume new responsibilities. This physical side of the choice of Superiors, and of those to be elected by members of the community, should always receive due attention, though sometimes it is entirely lost sight of. Not a few communities, however, have suffered in their usefulness and in the fulfilment of the design of their institute by the selection of Superiors whose neurotic conditions sometimes seemed to proclaim a high degree of piety, which was, however, rather emotional than practical. The physician's view of some of these cases would add materially to the knowledge of the character of such individuals.

It should in general be very clear that the development of any serious nervous disease, which is not likely to be cured by ordinary remedies or which requires freedom from responsibility as the first requisite for improvement, should be the signal for consideration as to a change of Superiors. Physicians see much more of the evil that may be worked in this way, and realise the true significance of what is often a sad state of affairs, much better than those who have not the secret of the cause of the unfortunate condition. It is almost needless to say that the question of obedience to some one whose responsibility is not complete, but is influenced by neurotic disturbance, becomes an extremely difficult problem for the subject, and one in which there is apt to be the feeling that it was not the original intention of his obligation of obedience to bind him under such circumstances.

With regard to women especially, it must be remembered that there is for them a period between the ages of forty and fifty, during which for several years they are extremely unsuited for the responsibilities and exacting duties of a Superior. These years prove even to mothers of families, surrounded only by their own children and the ordinary circumstances of home life, a time of worry and irritation that plays sad havoc even with the best of dispositions. Mothers constantly complain to their physicians of an irritability of temper which they can scarcely account for, and which makes them do and say things which they are extremely sorry for afterwards. It is easy to understand, then, that a Superior with still more insistent duties when brought in contact with a number of persons, some of whom are almost sure not to be entirely sympathetic, is likely to suffer from irritation that is not a sign of absence of a fitting religious disposition, but only a physical manifestation of the physical strain through which she has to pass at this time of life. The years of the menopause, to be very plain, should not be allowed to make a Superior's life miserable and to add to the difficulties that a religious community always has to face in its relations to its Superior and to one another. Charcot, the distinguished French neurologist, used to say that women should never be asked to assume special responsibilities during the days of their monthly period, for their judgments are often warped by their physical condition. It is doubtful whether, in the majority of normal women, this is quite true, though the expression deserves to be remembered. There is no doubt, however, that the years of the change of life do bring on very serious modifications of the character of the individual, and occasionally these changes are lasting.

JAMES J. WALSH.

EPILEPSY AND RESPONSIBILITY

From the very earliest times epilepsy has been looked upon as a mysterious and in many ways an inexplicable disease. The Romans spoke of it as the malum comitiale, the comitial disease, because if an attack of it occurred during the meeting of the Roman people known as the comitia, in which municipal officers were elected and other city business transacted, an adjournment was at once moved, and no further proceedings were considered valid. During more modern times, especially during the middle ages, and almost down to our own time, those affected by the disease frequently came to be looked upon as the subjects of possession by the devil. Hysterical manifestations were even more frequently considered signs of possession (diabolical manifestations) but even in our time it is not always easy to make the distinction between certain forms of hysteria and epilepsy. Many of these sufferers were considered as not responsible for their actions. In this respect, at least, the advance of modern medical science has only served to confirm the popular impression of less sophisticated times, and it has come to be recognised that quite a large number of the sufferers from epilepsy must be deemed lacking in responsibility.

There are few nervous diseases that have been more studied than epilepsy, and yet, because the ailment involves so intimately the relations of the nervous system and the bodily function, there are few diseases of which less definite opinions can be given. This is especially true as regards prognosis and the question of mental deterioration in any given case. As a matter of fact the extension of our knowledge of epilepsy, far from making the question of the responsibility of the epileptic under trying circumstances more easy of solution, has rather served to show how difficult this problem must ever remain.

There are many forms of the disease,—the frank epileptic convulsion in which patients fall down, are seized with certain convulsive movements, become pale and lose consciousness for a time and then come to with an intense feeling of weariness which usually prompts them to sleep for some hours—too familiar to need further description. There are forms of epilepsy, however, quite different from these. In some cases, the attacks occur only at night, and unless the patient happens to be watched for some reason, there may be no trace of their occurrence, except perhaps a sore tongue where it has been bitten, or an intense feeling of weariness and depression in the morning. In still other cases, the physical signs are lacking almost entirely. There may be only a momentary loss of consciousness. A distinguished professor of medicine in this country used to have a momentary attack of confusion, during which he lost the thread of his discourse, and always within a minute, with a somewhat flushed face, he was able to go on, though he had to begin with another idea. The so-called psychic epilepsy, in which the symptoms are entirely mental and consist of some marked change of disposition for a time, are now universally conceded as constituting well-marked phases of the disease. Curiously enough it is with regard to these obscure cases, uncomplicated by serious physical manifestations, that there is most mystery; and they seem to affect the mentality and to disturb volition and responsibility more than the supposedly severer forms which cause convulsive attacks and are so easy of recognition.

Certain forms of masked or psychic epilepsy constitute the most puzzling problem that the expert in nervous and mental disease has to deal with where criminal acts are performed, apparently without sufficient motive, and yet where the limits of responsibility must if possible be determined. It is easy to dismiss these cases and to consider that because a certain amount of intelligence has been displayed in the performance of the act, and because the patient ordinarily understands perfectly the distinction between good and evil. that therefore the will must have been entirely free in the accomplishment of the criminal action and the intellect must have understood what it was doing. As yet the general public refuses to take the standpoint of the expert in mental diseases in many of these cases; and only when clergymen also shall come to a realisation of the pathological elements undermining free will in these cases, that justice will be properly tempered, not by unworthy or misplaced charity, but by the mercy which, knowing all, has learned duly to appreciate what is and what is not criminal.

Epilepsy, in certain of its obscurer forms, is responsible for many conditions in which there is a sudden access of insane excitement of a violent, often very impulsive, character, though sometimes of very short duration. During this state the patient is practically irresponsible, and yet he may have sufficient control over his actions to enable him to work serious harm. Such a stage of excitement may last not more than an hour or two;

usually all trace of it passes off in a day or two; before and after it the patient may be in perfectly sound sense and in apparently good health. One of our best authorities here in America, Berkley, in his treatise on Mental Diseases, gives the following striking opinion on this subject.

"The subject of masked epilepsy and the consequent mania is replete with interest to the physician and the jurist, since such patients are prone to impulsive acts of violence and automatic states in which the most complicated, but entirely unconscious, actions and crimes may be carried out without premeditation on the part of the sufferer, being also out of all accord with his character during his intervals of mental health. Besides the irritability, impulsiveness is an equally characteristic feature. No form of insanity more frequently gives rise to assaults and murder than epilepsy, and in no form of alienation is the physician so frequently called to the witness stand to determine the responsibility of the criminal."

One of the most prominent features of all epilepsy is the well known tendency to irritability that characterises sufferers from the disease. This of itself is an index of the fact that their responsibility is somewhat lessened, since they are unable to withstand even the petty annoyances of life without exaggerated reaction. Friends of epileptics know very well that it is a preliminary symptom of the coming on of an attack of epilepsy for the patients to become even more irritable than usual. Just after the comatose condition which follows an attack of epilepsy patients are also prone to be very irritable. An attack of epilepsy is really an explosion of nerve force, for no rational purpose, along motor nerves. This same tendency to an unwarranted explosion of energy is liable to occur along other nerve tracts that rule the patient's disposition.

The main symptom of importance in the case, and the one on which depends the recognition of the existence of the epileptic condition, is the actual occurrence of typical epileptic seizures. These do not always occur. Sometimes the periodic attacks take the form of what are called epileptic equivalents, that is, certain anomalous states of consciousness or disposition, which can be accounted for only on the supposition that there is some more or less latent explosion of nerve force in progress. At times even so simple a condition as migraine so nearly simulates epilepsy of the psychical type, because of its complications and sequelae and the regularity with which it occurs, that it has been spoken of as an epileptic equivalent. There is no doubt that, in successive generations, epilepsy and migraine may have a relation to one another that is something more than merely a coincidence.

A very interesting feature of epilepsy for confessors and spiritual directors is the tendency to religious emotionalism which so often accompanies what is called idiopathic epilepsy. This means epilepsy that develops without a direct cause, and which is evidently dependent on some essential defect of the nervous system of the individual. In asylums epileptics that have become irrational are known for their religious manifestations, and very often for perversion of their religious tendencies. As has been well said, an epileptic may carry his Bible under his arm, read passage after passage from the Scriptures, sing psalms continuously, and yet be so ungovernable as to be a nuisance, and so irritable towards his fellow patients and attendants as to be a constant source of worriment. He may read just those passages which have reference to love and charity for one's neighbour and dwell on them until they become a bore by repetition, and yet in a moment of irritation implore to be allowed to get hold of some deadly weapon in order to kill the usually inoffensive person who has done him some imaginary injury.

This last is a marked feature of the disease, for epileptics are prone to foster fancied grudges, and to consider without due reason that they have been ill treated. This is especially true with regard to their relatives or to those in attendance on them, and must be always borne in mind when the subjects of epilepsy bring tales of woe and persecution, which they pour out to anyone who will listen to them, and especially to anyone whom they think will set them right. These fancied wrongs are as real to the patients themselves as if they had suffered from actual maltreatment. The idea of revenge may easily obtrude itself. It can be kept under control, as a rule, during ordinary health, between attacks, but just preceding or after an attack it may very well become of the imperative character that sets an uncontrollable impulse at work.

On the other hand, no class of patients is apt to exhibit the low cunning of the insane in so marked a degree as the epileptic. Not only this, but even during ordinary health between attacks they may, owing to their disposition, plan cunningly to simulate some of the symptoms of an attack and then accomplish a really malicious purpose with deliberation. In a word, these patients present to the alienist the most serious problem in the calculation of responsibility that can possibly be imagined. As an expert has declared, "It is ofttimes impossible to decide whether an assault has been committed with full consciousness, or in a transient but blind epileptic fury."

There are a series of attacks that occur in which there are some almost typical convulsive movements followed by loss of consciousness that simulate epilepsy very closely, yet are not true epilepsy. These attacks are usually due to some cerebral affection or perhaps to some injury of the brain. Chronic intoxications, that is, the long continued presence in the body in noxious quantities of some poisonous substance, are especially liable to cause these attacks, which are called from their character epileptiform. Characteristic epileptiform convulsions occur as the result of lead poisoning or from alcohol or syphilis. Lead poisoning, for instance, may very well occur in others than those engaged directly in the manufacture or handling of lead. Certain persons are extremely susceptible to the influence of lead. In them such small amounts as are contained in a hair-dye, or even in water that is being used by others without any bad effect, may cause particularly the nervous symptoms of lead poisoning.

Chronic alcoholism is also a relative term in this regard. Some persons are able to stand very large amounts of alcohol without serious consequences, even though it is taken for long periods. Others succumb to its influence very rapidly; some especially susceptible people are liable to suffer from epileptiform convulsions almost whenever they take alcohol to excess. This masked epilepsy may take on an anomalous form. The story is told of a student of a Catholic college in the eastern part of this country, who, during one vacation, was given as a joke by some friends a rather strong dose of liquor in a glass of ginger ale. He was very thirsty at the time and did not notice the presence of the alcohol until he had swallowed the whole glass. As he was well aware himself he was extremely susceptible to the influence of alcohol. During the course of half an hour he became almost wildly drunk, and going down the street with an open pocket-knife he murdered the first person whom he met, who happened to be an entire stranger to him. The occurrence took place in New Jersey, and, in spite of every influence that could be brought to bear—the incident took place some thirty years ago—Jersey justice would have its way and the young fellow of less than twenty was hanged.

The epileptiform attacks that occur in the midst of these intoxications are quite as likely to be accompanied by various forms of mental disturbance as are attacks of true epilepsy. Only one feature with regard to them is more favourable, and that is that the ultimate prognosis is not bad. The neutralisation of existing poison in the system, and the prevention of further ingestion of the toxic material, puts an end to the tendency to epileptiform convulsions, as a rule, and also to the mental symptoms associated with them.

Epilepsy remains, notwithstanding all the advance in modern nervous pathology, quite as mysterious a disease as it has ever been. It matters not what its cause, or how slight it may be, sooner or later it is almost sure to be followed by mental disturbance and deterioration of intellectual and will power. At times there are periodic attacks of mental perturbation that may become true insanity. Even the mild form of epilepsy known as Jacksonian epilepsy, and consisting not of general convulsive movements, but of convulsive movements in only one member or one side of the body, are, if allowed to continue, followed by some mental disturbance. It would seem as if the explosion of nerve force in the brain centres,—which, physiologically speaking, an attack of epilepsy evidently is,—causes eventual deterioration of the physical basis of mind and will, so that mental operations can no longer be performed with their wonted expertness or accuracy, nor decisions made as rationally as before.

In general, it is well understood that the more serious the epilepsy the more liability there is of the development of permanent mental disturbance. The earlier in life the epilepsy declares itself, too, the more unfavourable is the prognosis as to the enduring retention of complete mental sanity. In people in whom the epilepsy commences late in life, the process of mental deterioration does not begin to be noticeable so soon as when it occurs in younger years, and besides, it practically never runs a rapid course. Epilepsy, however, developing late in life, unless for some special cause, as injury or the development of syphilitic tumours in the brain, is an extremely rare affection. Idiopathic epilepsy, that is, epilepsy for which no definite cause can be discovered, is usually dependent on hereditary instability of the nervous system and is typically a disease of early years, of childhood and adolescence. According to the best authorities, about one-fourth of the cases of epilepsy make their appearance before the age of 7 years. Over 50 per centum of all cases develop before puberty. About one-third of all the cases develop between 14 and 20. And even of the remaining, less than 20 per centum, over 12 per centum develop between 20 and 25, leaving scarcely more than 5 per centum for all the remaining years of life.

Of course, even in severer forms of epilepsy, mental disturbances do not appear at once. It sometimes takes many years for the constantly recurring manifestation of explosive nerve force to produce the deterioration that gives rise to lowered rationality. Distinct mental deterioration is eventually inevitable, though modern experience with epileptic colonies, in which patients are enabled to live a quiet life, most of it in the open air and under conditions of nutrition and restfulness especially favourable for their physical well-being, shows that the development of insanity may be put off almost indefinitely.

There are many advertised cures for epilepsy. None of them is successful, and all of them may do harm. The bromides have a distinct effect in lessening the number and frequency of seizures, but if taken to excess they have a serious depressing effect upon the patient. There have been more cases of mental disturbance among epileptics, and intellectual degeneration sets in earlier, since the introduction of the bromides, than before. It is the abuse of the drug, however, not its use, that does harm. More important than any drug is the care of the patient's general health. The digestion must be kept without derangement; the bowels made regular; all sources of worry and emotional strain must be removed. Patients should as far as possible live in the country, and farm life has been found especially suitable. Relatives are often a source of irritation rather than consolation to these patients, and the life in epileptic colonies has been found eminently helpful.

JAMES J. WALSH.

PSYCHIC EPILEPSY

One of the most interesting phases of epilepsy is the type of the disease in which, without any significant motor symptoms, psychical manifestations prevail very markedly. A special manifestation in this affection is the occurrence of a more or less complete assertion of what is called a secondary personality. Apparently the individual becomes so divided in the use of the mental faculties that there are two states of consciousness. In one of these the patient knows and remembers all the ordinary acts of life, the other carries the record of only such actions as are done in a peculiarly morbid psychic or epileptic condition. It is rather easy to understand that this strange state of affairs may readily give rise to even serious complications as regards the individual's relations to others, and may make the problem of responsibility for apparently criminal acts that have been performed very difficult of solution. Undoubtedly, however, this set of phenomena constitutes a form of mental alienation that must be reckoned with in many more cases than might be thought possible. The difficulties that may have to be encountered in the proper appreciation of the actions of such individuals is best illustrated by some cases.

At a recent meeting of the New York State Medical Association a case was reported that shows how extremely difficult it may be to judge of responsibility under these pathological circumstances. The patient, a young man of about twenty-two, was the son of parents themselves of marked nervous heredity, signs of which appeared in other members of his generation. While in attendance at a public academy he had been quite severely maltreated during the course of an initiation into a secret society of the students—the more or less familiar processes known as hazing being employed. As a result of this he had suffered from an attack of unconsciousness that lasted for several hours. No other symptoms, however, or sequelae, appeared for nearly a year. Then, while boarding with his sister, he became morose and difficult to get along with. He quarrelled with his sister several times and generally their relations were rather strained. He came home one evening very late to supper, and because things were not to suit him on the table, he grew violently angry. He went upstairs to his room in this morose state and, procuring a revolver, after a short time came down and shot at his sister.

Fortunately he missed her. He at once left the house but was followed by his brother-in-law, and, after he began to run away, by others whose attention had been attracted by the shot. He left the country road and ran across the fields. He was found at the foot of a rather high stone wall in a state of unconsciousness. From this unconsciousness he did not recover until the next morning. In the meantime he had been brought home and put to bed. The next morning he claimed that he had absolutely no remembrance of anything that happened after he became angry at the table because of his supper. The family made no further difficulty about the matter, and, as nothing serious had resulted, the boy went home to live with his father on a farm and seemed to grow much more equable in temper.

One day, when very tired and out of sorts because things had not been going as he wanted them to, he was asked to clear a potato patch of potato bugs by spreading Paris green over it. Some hours later he was found in the field suffering from severe pains in the stomach and with evident signs of having swallowed some of the poison. A doctor was called, an emetic was given and he purged, and after a time he recovered from the symptoms of poisoning. He claimed that he had no recollection of what he had done, nor did he know how he came to take the poison. After this he begged the family to watch over him carefully and not to let him be alone at times when they recognised that he was somewhat morose in temper. He was not melancholic in the sense that he wanted to commit suicide, but something seemed to come over him in spells, and while in a state of mind of which he had no recollection afterwards, he performed actions that seemed voluntary and yet were not.

He did not have very good health on the farm, and so he was advised to try the effect of life at sea. A position as assistant steward was obtained for him on a coastwise vessel. In this position he gained rapidly in weight and seemed to have excellent health. All tendencies to moroseness of disposition disappeared. After a time he was promoted to a stewardship and later became the purser of a rather important vessel. He has given excellent satisfaction and feels in every way that he is in a much more balanced condition than ever before.

He still insists that he remembers nothing of how the two almost fatal incidents in his life came about. All his family are convinced that it was not a responsible state of mind that led him to attempt either of the crimes. It seems not improbable that this is one of those fortunately rare cases in which an attack of psychic epilepsy sometimes obliterates for a moment the individuality of a patient. At times these attacks last much longer, and the change to a secondary personality may represent a rather long interval. A number of cases of what are called ambulatory epilepsy have been brought to the attention of the general public of late years because of certain interesting features of the cases that have been exploited in the daily press.

Patients suffering from this form of nervous disease may wander from their homes, and while performing automatically a number of actions, such as buying tickets, travelling on cars and railroad trains, or even arranging the details of their journey for a long distance, may yet be in a state of mind that is not their ordinary consciousness. Men may leave home under the circumstances and find themselves after months in a strange town where they have established themselves in some quite different occupation from that to which they were formerly accustomed, or for which their early training fitted them. There seems to be an absolute division between the states of consciousness that rule the individual during the intervals of ordinary and extraordinary personality. There are, of course, many reasons for thinking that at times such a change of personality might be feigned; but many of the cases have been followed with too much care to allow this thought to serve as an explanation for all of them.

A case which serves to bring home very clearly the possibility of such a state of mind giving rise to serious complications is the following: The patient was a young man in attendance at the medical school of a university in a foreign city. He had been very careless in money matters, and had aroused family suspicion that even the money sent him for tuition was being used extravagantly. A friend of the family came to see him unexpectedly in order to assure himself how the boy was actually getting along. The boy's accounts were in a very disordered condition; he had not bought the books for which he pretended to want money; he had not paid his tuition. He realised that all this would come out as soon as the university authorities were consulted. Very naturally he was in an extremely perturbed state of mind.

While on the way to the university with this friend they passed a corner pharmacy, and the young man asked to be allowed to step in for a moment for a remedy for headache. The friend waited on the sidewalk for him, and when, after some minutes, the young man did not come out he went in to inquire for him, and found that after purchasing a headache powder the young man had gone out by a side door. For three days nothing was heard from him. Then a telegram announced that he was in a hospital in a distant city and that he had been picked up on the street unconscious. When he came to in the hospital he had no idea where he was, and, according to his own story, no recollection of how he got to the distant city.

It might be very easy to think, under such circumstances, that this was all pretence. A number of these cases of ambulatory epilepsy have been under the observation of distinguished neurologists, however, and there seems no good reason to doubt that some of them, at least, were entirely without any fictitious element. In any given case the possibility of the occurrence of an attack of what is really the assumption of a secondary personality must be judged from the circumstances, from the previous history of the individual, from the family traits, and from certain stigmata as narrowing of the field of vision and the like, which go to show the existence of a highly neurotic constitution. In this case the family history showed marked neurotic tendencies on both sides, and a brother had displayed a tendency to regularly spaced attacks of alcoholism about every six weeks, and finally became absolutely uncontrollable. There seemed good reason to think that the case was a real example of ambulatory epilepsy, and that the lapse of memory claimed by the patient really existed.

In these cases it is usual for the so-called secondary personality to assert itself at moments of intense excitement, especially if they have been preceded by days of worry and fatigue and nights of disturbed rest. The secondary personality is not a complete personality, but is a manifestation of the original ego with the memory for past events as a tabula rasa. It is well known that the memory is one of the intellectual faculties most dependent on physical conditions. It is the lowest in the scale of mental qualities and is shared to a very large degree by the animals. Injuries to the head not infrequently produce lacunae in the memory. These lacunae often have very striking limitations. It is not an unusual thing to find that old people remember events of their very early childhood better than things that have happened within a few years. Still more interesting is the fact that languages learned in youth may continue to be easily used, when those that were learned later in life, though perhaps known better than the previously studied languages, are forgotten.

It has often been noted that people who suffer from apoplexy may have peculiar affections of their memory. This may include such striking peculiarities as the forgetting of the uses of things, though their names are retained, or more commonly, the forgetting of names while the knowledge of uses remains. The one form of memory disturbance is called "Word Amnesia;" the other is called "Apraxia." It is on record that a person suffering from a hemorrhage in the brain has lost completely the use of a language acquired later in life, though the memory of the native language, long since fallen into disuse, was perfectly retained. One apoplectic woman patient who had left Germany before she was ten years of age, and who had lived in America until she was fifty, forgot absolutely the English she knew so well and had to set herself to work to learn it over again, though her German came back to her very naturally. These are wonderful peculiarities of memory-pathology that show how much this faculty is dependent on the physical basis of mind and upon the cellular constituents of the brain.

It is not surprising, then, to find that lapses of memory may occur and that, as a consequence, so many of the facts that ordinarily enable us to identify ourselves as particular persons may be in abeyance. That apparently a secondary personality asserts itself,—though not in the sense that there is ever another ego present, another mind or another will,—practically all experts in psychology and nervous diseases are now ready to concede. There are, however, involved in this question a number of important problems of responsibility that have not as yet been entirely worked out, and with regard to which prudent persons are withholding their judgment. Each case must be studied entirely on its own merits, with a leaning in favour of the criminal or patient, in case there are evidences in past life of serious disturbances of mentality, though only of very temporary nature, or if there is a strong nervous or mental heredity.

The notion of the possibility of a secondary personality asserting itself is a much older idea than it is usually thought to be. When Stevenson wrote Dr. Jekyll and Mr. Hyde, the immediate widespread popularity of the book was not due to recent psychological studies on dual personality and popular interest in a rare but striking mental phenomenon, but rather to the traditional feeling, long existent, of the possibility of two personalities in almost any individual. The other law in his members, of which St. Paul speaks, is an expression of this feeling, and its recognition was not original with him since it is after all a phenomenon at least as old as the existence of conscience. It is one of the basic ideas in religious feeling. Nearly everyone has something of the consciousness that there is in him possibilities for evil that somehow he escapes, and yet the escape is not entirely due to his own will power. There is here the mystery of temptation, of free will and of grace as the drama of conscience works itself out in every human being. At times the evil inclination seems to get beyond the power of the will and a period of irresponsibility sets in. Needless to say, the adjudication of how much may be due to the habitual neglect of repression of lower instincts is extremely difficult, and this constitutes the problem which the alienist must try to solve. In the meantime there is need in many mysterious cases where secondary personality may play a rôle, of the exercise of a larger Christian charity than that hitherto practised. Pretenders may succeed in deceiving only too often, but in the past not a few innocent individuals have been held to a responsibility for actions for which they were not quite accountable.

JAMES J. WALSH.

IMPULSE AND RESPONSIBILITY

Not unlike that condition which develops as the result of so-called psychic epilepsy, in which patients perform apparently voluntary acts, while the mind is really clouded by an epileptic attack, are those states in which, as the result of a more or less blind impulse, acts are performed for which the responsibility of the individual is at least dubious. Modern experts in nervous and mental diseases have sometimes spoken of these states as obsessions. This term is adopted from the older writers on mysticism who used it to designate states of mind in which an individual was under the influence of some spirit, though his intellectual and volitional state was not as completely under the subjection of this spirit as in the condition of possession.

It seems clear to the modern student of these obscure conditions that the old mystics and the modern alienists practically talk about the same state of affairs when using this term. As the result of obsession, mystical writers would have conceded that responsibility is not quite complete, though it is not entirely done away with. The modern alienist is just as sure of the diminution of responsibility, though he considers it due to the fact that for some physical reason the will is not able to act or prevent action as it is under normal conditions. The will is sometimes spoken of by certain of these modern psychologists as mainly an inhibitory faculty, that is, a faculty which prevents certain reflex acts from taking place, though permitting one set of reflexes to have its way. Under the influence of an obsession or, as the French call it, une idée obsédante, this inhibition is not exercised and as a result an action is accomplished which the agent may very shortly afterwards regret exceedingly.

There is no doubt that impulsions or impulsive ideas may push an individual into the performance of an action which his reason condemns. Uncontrollable anger is a well recognised example of this. Impulses of other kinds may exercise just as tyrannic a sway, though it is harder to recognise the elements that make up the mental condition in other cases. Of course it may well be said that man must control his impulses. It is, however, just such impulses as can not be controlled that lessen responsibility and sometimes seem entirely to destroy it. It would, without doubt, be very easy to advance the uncontrollable impulse as an excuse for many criminal actions. In fact, the discussion of responsibility and its limitation by impulse would seem to be open to so many abuses as to make it advisable, in the present indefinite state of our knowledge, to put the subject aside entirely. The argument, however, from the abuse of the thing, does not hold, and an effort must be made to get at the truth concerning certain mental conditions which modify responsibility.

It is generally conceded that no two men are free in quite the same way with regard to the actions which they may or may not perform. Allurements that are almost compelling for some individuals, for others have no influence at all. Some men are so under the influence of anger that irritation may easily lead them to the commission of acts for which they will be subsequently supremely sorry. This may even be the case to such an extent as to endanger their lives, yet they are not able to control themselves. Many men suffering from degeneration of the arteries of the heart have been warned, like John Hunter more than a century ago, of the extreme danger of a fit of anger, yet, like John Hunter, have succumbed to bursts of anger, notwithstanding the warning, because someone irritated them beyond their rather limited powers of endurance.

It is extremely difficult ever to come to any proper appreciation of the responsibility of a given individual from a single act. Preceding acts, however, may very well give evidence of the state of mind and the tendencies to disequilibrium which may make an apparently normal individual irresponsible under trying circumstances. The only way to render this clear is to illustrate such conditions by a concrete case.

Not many years ago one of the large cities of this country was shocked, for one twenty-four hours at least, by the news that a business man had shot his partners and himself, while at a consultation in which the affairs of the partnership were being settled up, after legal dissolution had taken place. The man in question had paid some debts of the firm with his own personal checks, and without taking proper legal recognisance for the moneys paid. When the partnership had been dissolved his partners insisted that instead of obtaining credit for these payments he should, on the contrary, pay his share of these debts once more as a partner. The state of the evidence was such that his lawyers told him it would be useless to take the case before the court at all; there was nothing to do but pay the unjust demands. He went to the meeting of his partners with a certified check for the amount of their claims in his pocket. As he took out his pocket-book to pass it over to them he seems to have realised very poignantly the fact that he was paying money that he knew he did not owe, and that his partners knew he did not owe, and that they were evidently taking advantage of a legal quibble in order to cheat him. Evidently it was an extremely trying situation. It was too much for his mental balance and he took a revolver from his pocket, shot both his partners dead, and then shot himself.

Taken by itself it is extremely difficult to say anything about the responsibility of a man who commits an act like this. In ordinary life he was known as a clever business man; to his friends he was known to be rather irascible and impatient, but a fairly good fellow. He was known to have what is called an awful temper; he had, however, never committed any violent act before. It is possible, of course, that a man should give way to a fit of anger for the beginning of which he is responsible, and then do violence much greater than he would justify himself for in calmer moments.

There was another occurrence in the man's life that seemed to throw informing light on his mental condition. When he first came to live in the large city in which he died he began paying attention to a young woman, and the young woman was informed by a friend that he probably had a wife living. The young woman investigated this by putting the question directly to him. He denied it at once, wanted to know the name of her informant, and finally laughed the whole matter out of her mind. Within a week after his marriage to her, while on their wedding tour, he was arrested, charged with bigamy at the instance of his first wife, and it became evident at once that the charge was well substantiated.

Here is a man, then, who twice at least in life, when put in the presence of trying conditions, goes on to do the irretrievable, though the act is eminently irrational.

With regard to the murder and suicide it is said that he had talked to friends of shooting the scoundrels who were cheating him, but had been persuaded of the utter foolishness of any such idea. He had apparently given it up entirely. Notwithstanding this, he went to the last conference with his former partners with a loaded revolver, as well as the certified check for the amount of their claim. In the case of his bigamous marriage, notwithstanding the warning that his second fiancee's questions must have been, he followed out his preconceived idea of marrying her, though he must have realised in saner moments that discovery of his double dealing was inevitable. In a word, he was a man who, becoming dominated by an idea, an obsession it may be called, to do something, could not get away from the sphere of its influence even though it might be made very clear to him it was eminently irrational to follow out the idea.

There are many such individuals, and only the knowledge of their previous career enables us to desume the responsibility for their acts under trying conditions. That they are not responsible in the ordinary sense in which the logical, timorous mortal is who is at once repelled from such modes of action seems very clear. Their lack of responsibility is manifest, at least to a degree that makes it easy for charity to find excuses for their crimes because of fatal flaws of character, the result of physical defects and faulty training, which make themselves felt especially at the moments that try men's souls.

JAMES J. WALSH.

CRIMINOLOGY

In recent years no little attention has been devoted to the subject of criminology, and a supposed science of the criminal has been evolved. It has been the claim of a very well known Italian school of mental diseases, whose leader is Professor Lombroso of Turin, that there is a criminal type in humanity, that is, that there is a generic human organisation not difficult of differentiation, at least as a class, the members of which almost necessarily develop criminal proclivities. Even when criminality has not actually occurred, this is thought to be but an accident, and criminal acts may be looked for at any time from these individuals. Lombroso's claims in this matter have met with decided opposition in every country of Europe and also here in America. This opposition has come especially from serious students of abnormal types who have devoted much time to the study of criminals and other supposedly degenerate individuals. Magnan, the very well and widely known French authority on insane peculiarities, especially the so-called criminal monomaniacs, and whose opportunities for careful investigation of such cases in the Asile St. Anne in Paris have been very extensive, utterly rejects the idea of a special physical conformation as characteristic of the criminal.

He is not the only one of the distinguished authorities in mental diseases who is in opposition to Lombroso in this matter. Dr. Emile Laurent, the eminent criminologist of Paris, has shown that the same anomalies which are supposed to characterise criminals are to be found among those who have never committed any criminal act, and that these supposed signs of degeneracy are not sufficient to indicate even that there are criminal tendencies. Manouvrier, the distinguished anthropological authority of the University of Paris, does not hesitate to advance the opinion that he can not find any distinctive difference between criminals and normal men in the extensive studies of the comparative anatomy of the two classes. He admits, however, that environment sometimes leads to the formation of habits which modify the anatomy in certain ways, and that of course traces of hard work, as well as of poor living, can be found in the anatomical conformation of many habitual criminals.

Dr. von Holder, a distinguished German authority on the subject, says that it is impossible to draw any conclusion from cranial asymmetries as to psychical characteristics, and that physical signs of degeneration indicate nothing further than the possible presence of a tendency to psychic degeneration. Dr. Wines, quoted by Draehms in his book on The Criminal, a Scientific Study, says that in a strictly scientific sense, the existence of an anthropological criminal type has not been proved, and it is doubtful whether it ever can be proved. Dr. Arthur McDonald, the well known American specialist in criminology and degeneracy, some of whose work in connection with the National Bureau of Education at Washington has attracted widespread attention, says, in his Abnormal Man: "The study of the criminal can also be the study of a normal man, for most criminals are so by occasion or accident, and differ in no essential respect from other men. Most human beings who are abnormal or defective in any way are much more like than unlike normal individuals."

How much the subject of criminology has been overdone because of the morbid popularity of the idea that many persons are, as it were, forced by their natures into the commission of crime, can best be appreciated from some recent publications with regard to left-handed individuals. A number of supposed observers, much more anxious, evidently, to make out a case for a pet preconceived theory, than to make observations that would add to the present store of truth, have rushed into print. As a result, left-handed persons have been said to be criminals much more commonly than those who habitually use their right hand, and have also been said to be defective in other ways. They were spoken of as weaklings, degenerates, and the like. Statistics even were quoted to show a much larger proportion of criminals than might be expected, according to the normal percentage, between right-handed and left-handed people, among those who use their left hand by preference. As a matter of fact, left-handed people are far from being the weaklings or degenerates they are thus proclaimed; but on the contrary are often magnificent athletes and excellent specimens of normal development. Left-handedness is due to right-brainedness and this is an accident dependent on a diversion of blood supply in an increased amount to this side of the brain in early embryonic life. This question of the criminal and the left-handed individual and their mutual relations is only a good example, then, of how far over zealous advocates of a theory have been led astray in their attempts to bolster it up.

Draehms, whose opinion on the supposed born criminal is worth while quoting, as it is founded on his personal experiences and observations while a resident chaplain of the state prison at San Quentin, California, says:

"Crime is not, as Lombroso and his coadjutors would have us believe, wholly either a disease or a neurosis in the sense of a direct, absolute, physiological, pathognomonic entity, though doubtless not infrequently closely associated with physical, anatomical, and nerve degeneration, as above conceded. To presuppose absolute and necessary brain lesion or diseased nerve action, or anomalous, physiognomonical, or anatomical diathesis, as the inevitable precursor of any form of mental and moral deflection, is an assumption wholly unwarranted and is nowhere substantiated by facts, though its advocates have sought to lay their foundations deep and wide in the materialistic hypothesis. Most criminals present unusually sound physiological conditions, and there is among them no unusual death rate, considering their habits and mode of life, as we shall hereafter see. Hence their moral instability can not be associated with physiological instability in the absolute sense. The physical defect must be either reversionary or incidental, rather than absolute."

The impetus in the study of criminals, which came as a result of the revolutionary teaching of the Italian school, has not been without a good effect. Criminals all over the world have been studied more closely and more sympathetically, in order to test the new ideas, until now it is possible to draw definite conclusions with regard to certain features of the problem. After a time, Lombroso came to admit that the so-called criminal type occurred in somewhat less than half the criminal cases. Criminal anthropologists, however, have shown that the physical conformation called by the name criminal, is really only the result of a defective or degenerative physical constitution. It is easy to understand that persons born with a defective nervous system, or with serious degenerative lesions in other parts of the body, which prevent the proper nutrition and functional development of the nervous system, would perform many more materially criminal acts than the rest of the population. The idiot and certain forms of the degenerative insane show this. Any defective development of the nervous system, moreover, may lead to instability of moral character, because the free action of the soul may be hampered by the physical environment with which it is associated.

Certain of the physical peculiarities most frequently seen in criminals have an influence of this kind and merit discussion. A knowledge of them will furnish clergymen with reasons for a larger charity to those unfortunates, and a greater tolerance for their relapses, without allowing sentiment to play too important a rôle in dealing with them. There are all grades of defective human beings, from the idiot up to those little less than normal. Anatomical peculiarities prevent the proper functions of the nervous system, as it is not hard to understand. The will is hampered in its action by the defect of the instrument through which it must work.

In persons properly to be considered as degenerates usually the head is small, though this may not be very noticeable because of over-development of the jaws. A heavy lower jaw particularly, because of the principle of bone-development that size depends on functional action and reaction, may lead to over-thickness of the skull at the point of articulation. The jaw articulates with the base of the skull, and as a consequence the cranial capacity of these individuals is distinctly less than normal. Besides this, there is commonly some abnormality in the shape of the head, or the cranium is distinctly asymmetrical. It has been noted that criminals have a large orbital capacity, that is to say, the bony framework surrounding the eye is so large as to encroach much more than usual upon the space left within the cranium for nervous tissue. The bones of the skull are likely to be thicker and heavier than usual, thus also limiting the cranial capacity. The superciliary ridges often project and give the beetling brow that is sometimes so remarkable. The jaws are heavy, and especially the lower jaw is apt to be large and prognathic, that is, projecting. This may extend even to the existence of a so-called lemurian appendix of the jaw. The zygomatic process is apt to be prominent, in keeping with the heaviness of the upper jaw. The nose is usually somewhat flattened, and may be crooked. This peculiar development of the nose puts most of the internal parts of that important organ within the skull itself. This further encroaches upon the cranial capacity. The ears are asymmetrical, often unevenly placed at the sides of the head, sometimes adherent at the lobule, sometimes very prominent. The displacement of the soft tissues is due to the existence of asymmetry of the skull. As may be seen, all of the characteristics of the criminal type, pointed out by Lombroso, may practically be summed up in the one expression, there is diminished amount of intracranial space.

With regard to many cranial deformities, and especially various thickenings of the cranial bones, it must not be forgotten that they are not the expression of physical heredity, but are often pathologically acquired. Certain diseases of children are accountable for many of them. Various disorders of nutrition in the early years of life express themselves in bony deformities, and the skull is not spared. Rickets, for instance, is well recognised as a cause of such deformities. Owing to a wrong etymology of this word, by which it is supposed to be derived from the Greek word , meaning the spine, rickets is sometimes scientifically called rachitis. The connection, then, between the cranial deformity and some underlying nervous disturbance might be assumed. It does not exist, however. Rickets is an English word, the derivation of which is unknown, but probably it is wricken, twisted, deformed, and its use has crept in because the disease was first described in England, and is indeed often spoken of on the continent of Europe as an English disease. Not that it is any more frequent in England, however, but was there first recognised as a distinct pathological entity. As the result of this affection the children, usually of poor parents, suffer from gastro-intestinal disorders of various kinds, and develop symptoms of malnutrition, affecting especially bone tissue. The ends of the long bones at the wrists and at the ankles, where the effects of the disease can be noticed particularly, become more thickened and nodular than usual. The ends of the ribs, where the bones join the cartilages, also become nodular, so that a series of beads can be seen down each of the child's sides, a condition described as the rickety rosary. In a similar way the bones of the skull become thickened, especially at the edges of the fontanels, that is, the openings in the child's head before complete ossification of the skull has taken place. As a consequence of this thickening these openings do not close as they should, and the head becomes markedly deformed in some cases.

Indeed, as has been shown by experts in children's diseases, many of the peculiarities that have been pointed out by over enthusiastic craniologists as indicating criminal degeneration, are really the results of the rickety process on the skull. Needless to say, however, this does not change the character of the individual, nor is there any good reason why such deformities should have any special connection with criminality. It happens that many of the criminal classes suffer from malnutrition in their early childhood, and as a consequence there is a faulty bony development of the skull. It is observations of this kind, particularly, that have served to discredit craniology as an independent science.

With regard to habitual criminals, the question of criminality must be discussed from the standpoint, not of those who theorise, but of those who know from actual observation most about the criminal classes. In an article in The Nineteenth Century and After for December, 1901, Sir Robert Anderson discusses how to put an end to professional crime. Sir Robert has been Chairman of the Criminal Investigation Committee of the English Parliament for many years. His opinion, then, is worth weighing well and is very strikingly different from those of the criminologists who would find a very large proportion of criminals among mankind. He says:

"I am not turning phrases about this matter, or dealing in rhetorical fireworks. I am speaking seriously and deliberately, and I appeal to all who have any confidence in my judgment and knowledge of the subject, to accept my assurance that if not 70,000 but 70 known criminals were put out of the way, the whole organisation of crime against property in England would be dislocated, and we should, not ten years hence but immediately, enjoy an amount of immunity from crimes of this kind that it might to-day seem Utopian to expect. The criminal statistics cult blinds its votaries. It is the crime committed by professional criminals that keeps the community in a state of siege. The professional criminals are few and I may add they are well known to the police. The theory that these men commit crimes under the overpowering pressure of habit, or of impulse, is altogether mistaken. They pursue a career of crime because, as Sir Alfred Wills expresses it, they calculate and accept its risks. And just in proportion as you increase the risks you will diminish the number of those who will face them. True it is that the army of crime includes a certain number of wretched creatures who have not sufficient moral stamina to resist the criminal impulse. I believe there are fewer of this class in England than abroad, but I know that these are not the sort of criminals whose crimes perplex the police. The high-class criminal is a different type of person altogether."

Sir Robert gives an extract from one of the morning papers of the day on which he wrote these lines, in order to show how different is the status of every ordinary habitual criminal from that which the enthusiastic criminologist supposes it to be:

"Hewson Patchett, 48, was sentenced yesterday for obtaining seven pounds and a gold watch by false pretenses. He urged it was his first offence, but a London detective informed the court that there were about two hundred cases against him for housebreaking."

Sir Robert adds: "If Patchett is a cool-headed, deliberate criminal, the whole proceeding is a farce. And if he be one of those miserable, weak creatures who can not abstain from crime, the sentence is barbarous."

Such experiences as Sir Robert hints at as occurring frequently in England, are certainly by no means uncommon in this country. Within the past year in at least four cases in New York City, in which a burglar, besides committing robbery, wounded or killed some one, either in the commission of the crime itself or in endeavouring to avoid arrest afterwards, there were more than two convictions registered against the criminal in his previous life. There can be no doubt that criminality becomes for some men a sort of mania, and that society must protect itself against their actions quite as it does against those of the insane by confining them under surveillance. It seems very clear that while a man may, under stress of circumstances or because of some specially tempting opportunity, be induced to commit burglary or some other crime by violence in order to obtain money, this will not happen a second time, except in the case of certain individuals whose moral tone is so perverted that reformation is practically hopeless. If a second conviction for burglary, therefore, is secured, a longer sentence than is now the custom should be inflicted, and the individual should not be allowed to go from under the surveillance of the authorities until he has demonstrated, for at least five years, his willingness and capacity to earn an honest living.

This may seem a drastic method. It may also appear to some that there would be consequent upon this system of regulating criminals a very undesirable increase of our present rather extensive system for the care of criminals. Here is where Sir Robert Anderson's experience is of value. The confirmed criminals are not near so many in number as is usually supposed, or as is even claimed by certain heedless statistical experts in criminology. There is no doubt, however, that these men succeed in drawing others around them and in organising most of the crimes of violence that are committed. There is a certain glamour about the successful burglar that allures the young man and starts him in the downward path of criminal tendencies from which he may not be able later easily to withdraw.

If those who are most deeply interested in the reform of the criminal classes would unite in an effort to secure legislation to the effect that the habitual criminal should receive, not a definite sentence but an indeterminate sentence; that is to say, that on his second conviction for burglary, he should be sent to jail until such a time as, in the opinion of officers of the institution where he has been confined, he shows signs of a disposition to become a worthy member of society, and that then he should be allowed to be at liberty only under such circumstances as would permit of reports with regard to his conduct for a time equal at least to the years spent in prison, then there would be much less need of the theoretical considerations with regard to the heredity of criminal traits, and the supposed all powerful influence of environment in fostering criminal tendencies. There is in this matter a very worthy field for the development of philanthropic qualities, and the student of the abnormal man will find many opportunities for the exercise of a large-hearted charity, rather than the facile condemnation which places all violations of law under the head of criminality.

Those who have made special studies with regard to criminals have, as a rule, come to the conclusion that our modern method of treating those convicted of crime is eminently irrational. It is a rare thing to pick up a newspaper without finding that a crime by violence has been committed by some one who has previously been in state's prison for a similar crime. Most of the burglars have a police record. Pickpockets and others continue to pursue their avocations, notwithstanding a series of convictions. It is clear that a sentence of a year or two, or even more each time that a crime is committed, does not act as a deterrent. Such people are differently constituted from those who are influenced by public conviction of crime and restraint of liberty. There is something radically wrong with their moral sense. It would seem that the proper way to treat them is after the same fashion as the method used with those who are mentally impaired.

After a man has shown, by a second conviction of a crime by violence, that he is one of those whose moral sense can not be restored by punishment to a realisation of his action, then an indeterminate sentence, somewhat as in the case of the mentally unstable, who are allowed to leave the asylum but are kept under observation, is the only proper method.

Men like Sir Robert Anderson are sure that this procedure could be adopted with regard to quite a liberal number of leading criminals whose influence induces others to crime. There would be much less need for all machinery of the criminal law than at the present time, and the community would be better protected. This is certainly true as regards the large cities, where crimes against property are almost without exception committed by those who have been previously convicted for such crimes, or who at least have been in intimate association with such convicted criminals.

This view of the criminal, as one against whom society must protect itself just as it does against the insane, is comparatively modern. It must be borne in mind, however, that insane asylums are by no means an old institution, and that the present restraint of very large numbers of the insane is something unknown before in history. It seems not unlikely that if this newer aspect of criminology could be made popular great benefit would follow, not only to the peace of the community and the freedom of its members from fear as to such crimes, but also a number of the weaker individuals, who are now influenced and led astray by clever criminals, would be saved from commission of crime and the necessity of punishment, with the degradation and lifelong stigma that this involves.

This is an aspect of criminology with which the Christian clergyman can be in sympathy, and that does not smack of the utter materialism which was at the foundation of much of the discussion of the so-called criminal type. The recognition of moral perversion as a form of insanity requiring treatment and then constant observation for many years, just as in the case of mental disequilibration would be a distinct advance over our present crude methods of dealing with criminals.

JAMES J. WALSH.

PARANOIA A STUDY IN CRANKS

Of late years the crank, in the various forms in which he or she may occur, has became a subject of great popular as well as scientific interest. As a matter of fact, the queernesses of people are a more absorbing study to the neurologists and psychologists than are any forms of insanity. It not infrequently happens that individuals of peculiar tendencies are prone to have special affinity for religious ideas, and strange applications of Christian formulae of thought. Even when they do not become absolutely insane in their religiosity, they may often go to extremes. It must be remembered, too, that some cranks are mentally affected in but mild form, and it may be difficult to determine whether their oddity is really the result of a certain amount of mental torsion, or merely intellectual tension.

Such persons are more likely to be brought in close contact with their pastors and other clergymen and with religious Superiors of various kinds than even normal individuals. They often put their confessors, particularly, in serious quandaries in the matter of spiritual advice. A review, then, of the accepted ideas of experts with regard to such people is likely to be of special service to those who would understand these cases as well as possible, though the present state of medical knowledge, here as elsewhere, leaves much to be desired.

A distinguished authority in mental diseases once said, half in jest though he meant it to be taken at least half in earnest, that a great many more of us are cracked than are usually thought to be, only that most of us succeed in concealing the crack quite well. The frequency of the crank adds to the interest of his study, which is by no means a department of medical science of recent growth. While interest in this class of persons has become much more intense in recent years, eccentric individuals have been an object of close observation and of serious study almost as far back as history goes. When Quintilian said that genius was not far separated from insanity, he meant to record the conclusion of his time, that men of genius are apt to seem inexplicable in their ways to those who come closely in contact with them. Eccentric persons, however, are by no means always undesirable members of a generation. It has been noted by historians in all ages that to the refusal of eccentric individuals,—often thought at the beginning, particularly, to be little better than insane—to accept the traditions of the past, we owe many of the privileges which we enjoy at the present time. Their refusal to think along old lines of thought often makes them valuable pioneers in progress.

Definite knowledge with regard to the pathological basis of crankism, or eccentricities, has not yet been obtained. What has been learned, however, has enabled the neurologist to distinguish various forms of mental perturbation, to recognise the probable influence of certain conditions and environments on the future action of eccentric individuals, and to foretell the probable outcome of the cases. All of this information is of very practical importance to religious Superiors and others in positions of religious confidence, who are sure to be brought, even more than the rest of the community, in contact with the eccentric class. It has seemed advisable, then, to condense some of the recent knowledge on this subject into popular form for the use of confessors, spiritual directors, and those in religious authority.

How recently medical knowledge on this important subject has developed along strictly scientific lines may perhaps best be appreciated from the fact that Professor Mendel of Berlin, to whom we owe the term paranoia, the recognised scientific designation for crankism, is yet alive and continuing his lectures on neurology at the great German university. The term, from the Greek word , meaning alongside of, and , mind, expresses the fact that the mental faculties of individuals designated by it are beside themselves, that is, the mental powers are not entirely under the control of the individual, so that they only come near voluntary intellection in its highest sense. In a word, the term contains a series of expressive innuendos by its etymological derivation.

Professor Berkley of Johns Hopkins University says that the word paranoia was first adapted by Mendel from the writings of Plato, to indicate an especial form of mental disease occurring in individuals capable of considerable education, at times of brilliant acquirements, yet possessing a mental twist that makes them a class apart from the great mass of humanity.

Professor Peterson, the President of the New York State Commission of Lunacy, gives a very good definition of the condition which, though couched in somewhat technical terms, furnishes the most definite idea of the essential properties of paranoia. He says: "Paranoia may be defined as a progressive psychosis founded on a hereditary basis, characterised by an early hypochondriacal stage, followed by a stage of systematisation of delusions of persecution, which are later transformed into systematised illusions of grandeur." He continues: "Though hallucinations, especially of hearing, are often present, the cardinal symptom is the elabourate system of fixed delusion."

In a word, the paranoiac is a crank usually descended from more or less cranky ancestors, with an overweening interest in his health to begin with, who later develops the idea that many people are trying to do him harm, or at least to prevent his rise in the world, and who finally becomes possessed of the notion that he is "somebody," even though those around him refuse to acknowledge it and pay very little attention to the claim. Such people not infrequently hear things that are not said. That is, not only do they hear the voices of the dead, of spirits good and evil, but also the voices of living persons, who are at a distance from them and sometimes even when those living persons are present, but have said absolutely nothing. These delusions of hearing, however, are not so important as the self-deception forced upon them by their mental state with regard to their importance in the world and their relations to other people.

The most significant consideration with regard to paranoia is the fact that it is practically always hereditary. Krafft-Ebing said that he never saw a case of true and reasonably well developed paranoia without hereditary taint. This does not imply, of course, that the same symptom of delusions exists in several generations, but some serious mental peculiarity is always found to exist in the preceding generation. Other authorities are not quite so sweeping in their assertion of heredity for these cases, though practically all are agreed that in over 80 per centum of the cases, some hereditary element can be traced without overstretching the details of family history that are given.

Paranoia occurs a little more commonly in females than in males. As it is of hereditary origin, it is not surprising to find that the peculiarities are noticed very early in life, though they may not be sufficiently emphasised to attract the attention of any but acute observers, who are brought closely in contact with the patients. Even in childhood, patients who subsequently develop serious forms of paranoia, usually have been shy, backward, inclined not to play readily, irritable, peculiar, precocious, prone to spend much time in study at an age when they ought to be interested mainly in sports, and they are generally old beyond their years. A typical example of this was Friederich Nietzsche, the German philosopher, who died a few years ago in an insane asylum.

Olla Hanssen, Nietzsche's biographer, who carefully collected the family accounts of the philosopher's childhood, said that he did not talk until much later in childhood than is usual. "As a boy he was retiring and solitary in his habits. During his school days he was always interested in books not in sports, in lonely walks not in young companions." A history of this kind will be found in the early career of many queer folk. Very often these old-fashioned traits are a source of pleasure to parents and sometimes even to teachers. During childhood, however, the sports of childhood should satisfy the child, and abnormalities of interest in things outside of childhood's sphere are always suspicious. The growing organism needs, first of all, muscular exercise, and after that the freedom of mind that comes with spontaneous play. It may be said, in passing, that the walk of a city child with its maid, when even the child's choice in the matter of where it shall walk is not consulted and the maid's will is constantly imposed, is the worst possible training for spontaneous action or volition in later life.

In the cases that develop early in life it will practically always be found that infantile cerebral disorders of some kind are a prominent feature of the history. The mother's delivery was difficult perhaps, and the child was for some time after birth unconscious, or infantile convulsions occurred. It may be remarked here that a history of convulsions in childhood is now considered by physicians as of serious import for the future nervous and mental life of the child. It has recently been announced, for instance, that so-called idiopathic epilepsy,—that is, epilepsy without some directly immediate cause,—very seldom develops later in life in persons who have not had in childhood convulsive seizures as the result of some extreme irritation. This does not imply that every child that has convulsions will suffer from some serious nervous or mental condition later; but every child whose mental and nervous equilibrium is not stable, because of hereditary elements of weakness, will almost certainly suffer. Injuries to the head in childhood are nearly of as great importance as the actual occurrence of convulsions.

There are usually three stages of paranoia described by authorities in mental diseases. These have been called the prodromal or initial period, which is also, because of the set of symptoms usually most prominent in it, often called the hypochondriacal stage of the disease. The patient occupies himself with his feelings and his sensations. He is concerned very much about the state of his health and is prone to think himself affected by diseases that he reads about or hears described. This set of symptoms, by itself alone, is not an index of enduring mental disturbance, but may be only a manifestation of a passing mental perturbation in sympathy with some slight physical ailment. This state may indeed be nothing more than the result of too persistent introspection. Most medical students suffer from a certain amount of hypochondria during their early acquisition of a knowledge of the symptoms of disease.

In the true hypochondriac, however, every bodily sensation, or as it is technically called, somaesthetic sensation, is translated to mean a significant symptom of serious disease. A slight feeling of fatigue becomes to the patient's mind the "tired feeling" of a dangerous constitutional disorder.

Any peculiar feeling, such as that of the hand or foot going to sleep, is set down at once as a symptom of a serious nervous disease, or if the patient has heard that in old people numbness of the extremities is a forerunner of apoplexy, he is sure to conclude that apoplexy is threatening in his own case. Subjective sensations of heat and cold set him to taking his temperature and his pulse, and even slight variations in these are magnified into important physical signs of disease.

Very often such slight symptoms as passing lapses of memory are magnified into approaching complete failure of memory, and lassitude becomes a permanent loss of will power, evidently due to disease in the patient's mind, and there begins the persuasion that nothing can overcome it. Morbid introspection becomes, after a time, the favourite occupation, and every slightest sensation or feeling sets up trains of thought that lead to far-reaching conclusions with regard to physical weakness. The patient is apt to be greatly preoccupied with himself, to neglect his duty towards others, to be utterly selfish, to fail to realise how much sympathy is being wasted on him.

Some people never pass beyond this preliminary stage of the mental disorder. Usually, however, after a time the patient misinterprets not only his own sensations, but the actions of other people in his regard; he becomes suspicious and distrustful of everybody around him, sometimes even of his best friends. He is passing on to the second stage of the disease, in which he is sure to feel that he is the object of persecution. Just as he misunderstood his physical symptoms, so he misconstrues the actions of his friends. He is sure that they look at him curiously, that they smile ironically. Sometimes he thinks that they wink at one another with regard to him, or make signs behind his back that are meant to be derisive. Even harmless passing observations may be morbidly perverted into severe and inimical criticism of himself and his actions.

The paranoiac is now apt to enter fully upon the second or persecutory stage of his mental disorder. His distress and discomfort he attributes to those around him and he is sure that he is the subject of persecution. At first his persecutors are not very definitely recognised. No particular person is picked out and even no particular set of persons. There is just a vague sense of persecution. A distinguished neurologist once said that no sane person in this world, outside of a novel or a play, has time to make it his business to persecute anyone else. When people come, then, with stories of persecution, either they themselves are not in their right mind and are deluded as to the source of the persecution, or else their persecutor is not in his right mind and the case needs seeing to from the other standpoint.

After a time, longer or shorter in individual cases, the paranoiac begins to recognise definitely who his persecutors are. As a rule, it is not some single individual, but a combination of individuals. Already there is the beginning of the third state of the disease—the grandiose stage of the disease, in which the patient feels an extreme sense of his own importance. It would be derogatory to his self conceit to consider himself the subject of persecution by an individual, and so it is usually some society, or the government, or its officials, or some secret organisation that is persecuting him, and perhaps also persecuting those who are near and dear to him.

Sometimes it is the Odd Fellows, or perhaps the Masons, who are the persecutors. If the newspapers have recently had some account of the disappearance of Morgan years ago, and this subject crops up periodically in the papers, then the Masons become a favourite subject for paranoiacs' delusions of persecutions. Just after the Cronin murder in Chicago, the Clan-na-Gael became an extremely fearsome spectre for paranoiacs who thought themselves persecuted. It is of some importance to know, as a rule, what the usual reading matter of a patient is, and what things are likely in his past history to have impressed him, in order to realise what the real source of his delusions of persecutions are.

It is curious how rational these patients may be on all other subjects except the special topic of their delusion. During the past year a paranoiac has been under observation, who is considered a reasonably rational individual by those who know him well, who follows his daily occupation, that of clerk, without intermission and with business ability, who is a faithful attendant at church, and who is very kind to his family, but who is sure that he is the subject of persecution by the Clan-na-Gael. He never belonged to the organisation. He is not able to give any good reason why he should be persecuted, except perhaps the fact that, though an Irishman, he never did belong to them. He is perfectly sure, however, that they are planning to poison him and his family. He finds peculiar tastes in the tea and the coffee at times. He throws out these materials and insists on his wife getting others at another grocery store. He sometimes brings groceries home from a distance and yet finds that if he ever buys materials a second time in the same place, they are sure to have been tampered with in the meantime by emissaries of this secret organisation. He feels sure that he has seen these secret agents, but he is only able to give vague descriptions. Not a little of the prejudice against these organisations is really founded on such morbid suspicions.

Another form that the idea of persecution sometimes takes, in this second stage, is the delusion that the patient is neglected by those who should specially care for him or her. A woman insists that she is neglected by her husband. She may become intensely jealous of him and make life extremely miserable for him without there being any good reason for her jealousy. These cases are not nearly so rare as might be thought. On the other hand, men suspect their wives of unfaithfulness. This suspicion may go to very serious lengths in persecution at home, though the man all the time keeps his suspicions to himself, in order not to make a laughing stock of himself outside of the house. It is this curious mixture of rationality and delusion that is the characteristic feature of the disease. It is for this reason that these conditions were sometimes called monomanias, as if patients were really disturbed only on one point. As a matter of fact, however, patients are mentally wrong on a number of points, though there is some one mental aberration so much more prominent than other peculiarities that it overshadows the others.

It is not long after the persecutory stage sets in before patients are apt to become themselves persecutors of others as a result of their belief that they are being persecuted. The French have a suggestive expression for this. It is persécutés persécuteurs, that is to say, "persecuted persecutors,"—patients who are trying to repay supposed persecutors by persecution on their own part. Such patients, of course, very easily become dangerous. They need to be carefully watched. As a rule, the persons whom they are prone to select as the persecutors upon whom they must avenge themselves are absolutely innocent parties. At times they are even dear and well meaning friends.

After the persecutory stage in paranoia, comes the third, or so-called expansive period of the disease. It has been remarked that sometimes this develops as a sort of logical sequence from the patient's ideas of persecution. If he has too many enemies and if important secret organisations are trying to be rid of him, he must be a person of some importance. As a consequence he evolves for himself a royal or aristocratic descent, or hints that he is the unacknowledged son of great personages. In a kingdom royalty is, of course, a dominant idea. In a republic like our own, he may consider himself to be the President or the politician to whom the President owes his office.

Paranoia Religiosa.—Not infrequently the first hint of their supposed greatness comes to such patients suddenly in a dream or in a vision; when their expansiveness takes a religious turn, this is especially apt to be the case. They may believe themselves to be especially chosen by the Almighty, a new Messiah or even Christ Himself, come once more to earth. Such people may retain much of their rationality on most of the points relating to practical life, and yet have this hallucination as to their close relationship with the divinity. Not only may they retain their mental equilibrium on other points, but they may even give decided manifestations of great genius. This is, I suppose, one of the most interesting features of this form of mental disease, but it is well illustrated in the lives of many modern founders of religious sects, even in our own generation.

Such religious reformers as Mahomet and Swedenborg seem undoubtedly to have been afflicted with this third stage of expansive paranoia. In our own day there is no doubt that many of the founders of new religious sects, many of the heaven-sent apostles or reincarnations of patriarchs and prophets, the miraculous healers and the like, are afflicted in this same way. It is useless and entirely contrary to the known facts to put such people aside as mere imposters. Imposters they are, but they have imposed on themselves as well as on their followers. They are sincere as far as they go, and the mental twist that gives them their power has occurred in the midst of the manifestation of the intellectual faculties of a highly practical character and of executive ability, with wonderful capacity for the direction of complex affairs. A prominent neurologist said, not long ago, that the most interesting feature of Christian Science is to contemplate in the study of the movement how near people may come to insanity and yet retain their faculty to make and handle money and even accumulate fortunes.

Paranoia Erotica.—After the paranoia religiosa, the most common form of the disease is the paranoia erotica. There are authorities in mental diseases who do not hesitate to say that an excess of religiosity and of erotic sentimentality are more or less interchangeable. This declaration represents, however, the unconscious exaggeration of a mind unsympathetic towards religious ideas. But it must not be lost sight of that the two forms of excesses, erotic and religious, are more nearly related than would be ordinarily supposed, and that erotic manifestations may be confidently looked for in patients who have been afflicted by a too highly wrought religious sentimentality. St. Theresa seems to have realised this very well and has touched on the subject in one of her letters.

Ordinarily erotic paranoia manifests itself by the patient imagining himself or herself to be beloved by some one of superior station. This love is of rather a platonic character and the "lover" cranks are prone to pick out as the object of their attention and annoyance some young woman rather prominently in the public eye, but whose reputation is of the very highest. Mary Anderson was the subject of a good deal of this sort of persecution. At the present moment the well and favourably known daughter of a great millionaire is the subject of many such attentions.

These paranoiacs are apt to become dangerous if they are prevented from paying what they consider suitable attention to the object of their affection. In hospitals they have to be carefully watched, and more than one accident has taken place as the result of relaxation of vigilance on the part of their attendants. If kept from the object of his affection, delusions of persecution become prominent in the amorous paranoiac, and he may become a persecutor in turn and thus a dangerous lunatic. He can not be made to understand that the sending of flowers and photographs and letters is entirely distasteful to the chosen one. Fortunately, after a time, in many of these cases, a state of dementia sets in, and then the patients become mild-mannered imbeciles whom it is not at all difficult to manage.

As a rule where the patient has passed through the various stages of paranoia, dementia, with symptoms of imbecility, closes the scene. The paranoia may not always follow the course mapped out for it. Stages may be skipped, several forms of delusions may become prominent in the life of the individual at about the same time. The main feature of the disease is its progressive character, and its diagnosis depends on the queerness exhibited all during the course of life, as well as on the presence of hereditary neurotic influences.

Special Forms of Paranoia.—There are besides the two types described a number of special forms of paranoia, some of which aroused attention first under the form of monomanias, that seem to merit brief treatment by themselves. In their extreme forms they are easy of recognition. Milder types, however, may easily escape classification under the head of paranoia, because they are considered to be individual oddities and not due to any physical or mental incapacity. Undoubtedly, however, the study of these peculiar "types," as the French call them, from the standard of the alienist or expert in mental diseases, will serve to make clearer the real significance of many otherwise almost unaccountable actions. There is no doubt, that the consideration of these eccentrics as paranoiacs makes the charitable judgment of many of their acts much easier, and at the same time is of service in managing them. They are likely to be of much less harm to the community and to their friends, when it is realised that they are not to be taken too seriously, but that, on the contrary, there is ample justification for a benevolent combination of interests to keep them from injuring themselves and their friends.

Paranoia Querulans.—One of the most important and familiar forms of the special types of paranoia is what is known scientifically paranoia querulans, that is, the peculiarity of those who insist on going to law whenever there is the slightest pretext. It is pretty generally recognised that a goodly proportion of the civil suits that crowd our law courts are due to the peculiarities of these people who insist on having their rights, or what they think their rights, vindicated for them by a court of justice. There are very peculiar characters in this line, some of whom make themselves very much feared and detested by their neighbours. There are some individuals to whom the slightest injury or show of injury means an immediate appeal to the law.

Not infrequently these patients, for such they are in the highest sense of that word, waste their own substance and even the means of support of wife and children, on their foolish law schemes. When their queerness reaches a certain excessive degree its pathological character is readily recognised. In a less degree paranoia querulans may be a source of very serious discomfort to friends and neighbours without exciting a suspicion of its basis in mental abnormality. Not infrequently such patients become more irrational at times when their physical condition is lower than normal, and a return to their ordinary health makes them more amenable to reason and less prone to appeal to expensive litigation.

It is evident that the irrationality of frequent appeals to expensive and bothersome litigation should arouse suspicion. Such patients need to be cared for quite as effectually as those who have tendencies to gamble away their substance or to waste it in the midst of inebriety. Unfortunately it is extremely difficult to frame laws so as to meet such conditions. Severer forms of the affliction are readily recognised and the sufferer is properly restrained. I remember once seeing a patient in Professor Flechsig's clinic in Leipzig, who had been sent to the asylum because of his tendency to go to law on the smallest possible pretext. This patient's incarceration in the asylum was due to a very striking manifestation of his paranoia querulans. He answered an advertisement for a clerk, published by one of the large commercial houses. He found himself one of a row of applicants for the position, and as the member of the firm whose duty it was to engage the clerk was at the moment busy, he had to wait several hours before his application was heard and refused. He tried to secure a warrant for the firm in order to have them indemnify him for the time he had spent while waiting for his application to be heard, at the rate of wages they would have been bound to pay him had he obtained the vacant clerkship; only as they had spoiled a day he claimed a full day's wages.

This patient had been in the asylum several times before because of his tendency to go to law. He always gained in weight while in the asylum, became much more tractable and less querulous as his physical condition improved, and usually after some months could be allowed to leave the institution. He was, however, one of the inept. With the help of asylum influence he usually obtained some occupation more or less suitable, but was not able to retain it for long. When out of a situation he worried about himself, usually did not take proper food, and then soon his litigious peculiarities began to manifest themselves once more in such form that if he could get the money to retain an attorney, or if he could persuade one to take his case on a contingent fee, and he was very ingenious at this, he soon became a veritable nuisance to those around him. When in poor health he was never contented unless he had at least one lawsuit on his hands, and only really happy when he had several.

The Gambler Paranoiac.—A form of paranoia that inflicts almost more of human suffering on the friends of the patient than any other is that in which the sufferer is possessed of the idea that he can, by luck or by ingenious combinations, succeed in winning money at gambling. Milder forms of this paranoia are so common that it is the custom not to think of even the severer forms of the gambling mania as a manifestation of irrational mentality. When a man thinks, however, that he can beat a gambler at his own game, or when by long lucubration he comes to the conclusion that he has invented a system by which he can beat a roulette wheel, he is, on this subject at least, as little responsible as the man who thinks that he has discovered perpetual motion.

This form of paranoia inflicts suffering mainly on the near relatives of the patient. There is no doubt that when extreme manifestation of the gambling mania becomes evident, patients should be legally restrained from further foolishness. One difficulty with regard to the proper appreciation of gambling has been an unfortunate tendency to class gambling among the malicious actions. There are many people for whom only two sins seem to have any special importance, drunkenness and gambling. As a rule, there is not the least spirit of malice in the ordinary gambler; not meaning, of course, by this the sharpers, who try to make money at the expense of others, but the man who believes that, somehow, chance and fate are going to conspire to enrich him at the expense of others, though it must be confessed that he does not usually even think of this latter part of the proposition which he accepts so readily.

We have had in recent times so many manifestations of the practical universality of the gambling spirit, the belief by people that brokers and banking concerns are ready to make them rich quick, that we have in it perhaps the best illustration of the partial truth of the proposition that "half the world is off, and the other half not quite on."

The "Phobias."—Sometimes the special form of queerness takes on a very harmless aspect. Patients are worried because of the fact that they can not keep themselves clean. They want to wash their hands every time they touch any object that has been handled by others, whether that object seems to be specially dirty or not. Such patients may wear the skin off their hands washing them forty or fifty times a day. They almost absolutely refuse to touch a door-knob, because it is handled by so many people. They will consent to take only perfectly new bills. It is almost amusing to see the efforts they make to avoid shaking hands with people, without giving direct offense. When it comes to shaking hands with their physician, they are apt candidly to declare that he must not ask them to do so, because they can not overcome their feelings as to the possibility of contamination from hands that come in contact with so many patients. This fear of dirt has received the name Misophobia.

There are a number of other "phobias," and the patient's fears are manifested at the most peculiar objects. Agoraphobia, for instance, is the fear of crossing an open place. These patients begin to tremble as soon as they get away from the line of buildings in a street, in their way across the square. This trembling becomes actual staggering, with a sense of oppression over the heart that makes locomotion almost or quite impossible. Claustrophobia, the opposite of Agoraphobia, is the fear of narrow places, and prevents some people from going through a narrow street with high buildings. Many of these "phobias" have a physical basis in some organic or nervous heart affection.

The Tramp.—One of the striking manifestations of paranoia in our modern life is the tramp. Most people are inclined to consider that the cause for the wandering life of these unfortunates is rather what a distinguished physician euphemistically called by the scientific name, pigritia indurata, that is, chronic laziness, than any pathological condition of mind. Most tramps, however, will be found, on that close acquaintanceship which alone will justify judgment of their actions, to have many other peculiarities of mind besides the shiftlessness which prompts them to wander more or less aimlessly from place to place. After all, it will hardly be denied that the calm acceptance of the notion that it is more satisfactory to indulge in laziness and wander without home or fireside, suffering the many privations and hardships, especially from the weather which these creatures do, rather than work and be respected and comfortable among their fellows, is of itself irrational.

Many of these tramps prove on close acquaintance to be interesting pathological characters. Various stages of outspoken paranoia will be found to exist among them. It is not unusual to find that certain among them have acquired the idea not so uncommon now among large classes of humanity, that the world is so unjust in its treatment of the labouring man, that work seems to them almost a persecution that must be undergone for the sake of the pittance derived from it. Sometimes there is the actual extrinsic idea of personal persecution for some fancied wrong done to a large corporation during a strike, or labour troubles, which they cherish as the reason for which they have had to give up a fixed habitation, and resign the idea of supporting themselves honestly and respectably. This persecution stage of paranoia easily turns to the second phase of this affection as already described, that in which the fancied victim of persecution becomes in turn the persecutor. Tramps thus readily give way to even organised attempts at revenge upon social order, and are led to believe themselves justified in attempts to burn and otherwise destroy property because of their enmity towards property holders and employers generally. Not infrequently the third stage of paranoia, in which there are delusions of grandeur, may be observed.

Personally I have known two tramps who wandered about the country with these grandiose ideas. One of them thought that he had in his possession immense wealth in the shape of large checks, signed supposedly by various important capitalists, and even foreign rulers. These checks were actually signed in the names of these personages, at the tramp's own request, by any chance passer-by or acquaintance. This patient died in a country insane asylum in the demented stage of paranoia, having gone through all the usual phases of the disease. Another tramp was confident that each recurring election he was to be elected to one of the highest offices in the state, or even to be made President of the United States. Not every one was taken into his confidence in this matter, however. The simplest declaration after the election from any chance acquaintance would assure him of his success at the polls, and on more than one occasion he turned up at the Capitol to claim exalted office, but was generally inoffensive in his ways, and was rather readily persuaded that his term of office did not begin for some time. It is easy to understand that a person might come into the possession of the idea that the official holding office in his stead should be removed; the result might very well be one of the sad tragedies supposedly due to anarchism, but really to paranoia.

Of course as with criminals, so with tramps; not a few of them take up this manner of life without any sufficient justification in their mental state to lessen our worst opinion of them. I do not think I should hesitate to say, however, that the majority of these unfortunates present distinct signs of physical and mental degeneration and are rather deserving of pity and care than of condemnation. They need, as a rule, very special environment to enable them to lead ordinary, respectable lives, because they were not originally endowed with sufficient initiative and independence of spirit to enable them to carry on the struggle for life in the midst of the hurry and bustle of our modern civilisation. As the pressure of the time becomes severer, more of these unfit come into evidence. They arc examples of the lowered mental states, unable to stand the rivalry with fellowmen, and ready to give up the struggle whenever the example of others who have already given it up is brought prominently to their notice.

It is not a little surprising how many of these tramps belonged originally to excellent, respectable families. Careful investigation of their personal history, however, will show that they have been, as a rule, backward children at school, always a little awkward in the way they took hold of things early in life, unsuccessful in the rivalries of school competitions, and in their first efforts at labour after school days were over. They always needed the encouragement of those whom they loved and respected, to keep them at their unsatisfactorily fulfilled tasks. They were the predestined failures in life, and have found out their uselessness early in their careers. This is the view of tramp life that is coming to be realised as true by all those who have studied the question, not from the standpoint of theory, but of practical experience with it.

So-called Monomania. The old term for paranoia employed for a long time was monomania, a word coined by Esquirol at the beginning of the nineteenth century. This word has dropped out of the terminology of mental diseases because there is no such thing as a patient suffering from a single symptom of mental disturbance, that is, being mentally perturbed on but one line of thought. There are always others, though they may be hidden except from the careful investigator. When Esquirol introduced the term he applied it to the most prominent symptom of the patient's mental alienation, but did not intend it to be taken as excluding other symptoms by which the essential nature of the patient's insanity could be diagnosed. Careful study will always disclose the fact that other symptoms are present. The word monomania has been an unfortunate one for scientific psychiatry, because it has been abused to shield criminals. The plea is often heard that a person under charge of crime is really subject to some mania that brought about the commission of the crime.

We often hear of kleptomania as a defence for persons who have failed to recognise the distinction between meum and tuum, and are haled before the court because of the detection of infringements of this distinction. True kleptomaniacs there are, but there are always other symptoms of their mental disturbance besides the tendency to steal. Their queerness in other ways has usually been recognised by their friends and by their family physician before the incident which calls attention to this special form of disequilibration occurs. Kleptomaniacs, too, are usually prone to take things of little value, or not especially suited to their wants and for which they have practically no use.

It is true collectors, that is, those who have a hobby for gathering curiosities of one kind or another to make a collection, may become so interested in additions to their collection as to be tempted to appropriate to themselves articles of which they can not otherwise obtain possession. Such actions may easily go beyond the bounds of reason. It must be remembered however, that the collection mania itself is often so pronounced as to be a little beyond the bounds of ordinary rationality.

Other so-called monomaniacs have the same characteristic and are associated with related symptoms of mental disturbance. Pyromania is sometimes pleaded as a defense for arson. It is a legitimate defense, however, only when the careful tracing of the patient's history beforehand shows the existence of other symptoms of mental unbalance. The homicidal mania is of the same order. There have been cases where men seem to have delighted in inflicting injuries or death upon fellow creatures from pure malice. Such cases as that of Jack the Ripper, for instance, are undoubtedly due to a special tendency to take life. In these cases, however, associated symptoms are never lacking. It is not improbable that in Jack the Ripper's case a sexual element was present, because the victims were always of one low class, and that the general character of the murderer would have revealed his irresponsibility. There are several stories of children—whose mothers delighted in seeing their husbands, who were butchers, slaughter animals—who seem to have had a veritable mania for seeing blood flow and to have exercised it in the murder of human beings.

Only the most careful examinations of the previous life of the patient, the investigation of childish tendencies and habits at school, and the incidents of boyhood and youth will sometimes enable the expert to recognise the constant existence of symptoms of mental disequilibration, the decided manifestation of which leads to serious events in after life. Monomania as a defense for crime has brought expert evidence into great disrepute. In this matter the greatest care is undoubtedly needed, however, for it is easy to do great wrong and punish the irresponsible victim of an impulse over which he has no proper control. On the other hand, it must not be forgotten that no such thing is known to exist as the perversion of the will on a single point. Moral insanity with regard to one special set of actions is a delusion that the increase of knowledge with regard to mental diseases has erased from the text books on this subject.

Responsibility of Paranoiacs.—From what has been said it is easy to understand how difficult is the determination of the responsibility of paranoiacs. Many classes of persons ordinarily considered to be quite responsible for their actions are yet so circumstanced that they are led into the performance of actions usually not considered rational, though not tempted thereto by any benefit directly accruing to themselves. On the contrary, it not infrequently happens that the mode of life adopted by the paranoiacs is of such a kind as would of itself, because of the hardships involved or the mental trials, deter ordinary people from following it. Paranoia, at least in its severer forms, completely justifies the plea of irresponsibility for actions committed. When it is remembered, however, that paranoiacs are often cunning enough to take advantage of their own supposed queerness voluntarily to commit crimes they might otherwise be deterred from by fear of punishment, some idea of the difficulty of the decision in these cases may be appreciated.

It is important, of course, that the physician should, as far as possible, avoid falling into the error of judging such people too harshly, since after all on him depends the attitude of society towards them. It would seem to be quite as important that the clergyman should occupy an advanced position in this matter. It might seem that charity could easily be overdone; it must never be forgotten, however, that it is better that ninety-nine guilty should escape rather than that one innocent person should be punished.

As a matter of fact, prejudice is much more likely to be against the supposed criminal than in his favour. While it is often declared that too many persons, who have done at least material wrong, are allowed to escape deserved punishment, as our knowledge of mental diseases increases there is more and more of a tendency on the part of experts to recognise that for many apparently voluntary actions men have only a modicum of responsibility. Responsibility is, after all, not the same in all men, but modified very much by the character of the individual, by his environment and by the motives which have come to be the well-springs of his actions.

No two men are equal in their responsibility when there is question of certain temptations to do wrong. Some men find it perfectly easy to resist allurements to dishonesty which others can not resist. Some men are perfectly free as regards their attitude towards indulgence in spirituous liquors. Others find it almost impossible to resist their cravings in this direction. One might go through the list of passionate actions and find this true with regard to every one of them. If this must be admitted with regard to men who are considered perfectly sane and responsible, how much more so does it become true of those who are already somewhat mentally unbalanced?

Unfortunately, the tendency to judge harshly, rather than mercifully, still continues to be one main reason for the infliction of punishment where often it is not deserved. Above all the clergyman must be a leader in this tendency to mercy, and his influence should be felt in popular education in this regard. It only too often happens that clergymen are found to be strenuous upholders of the opinion that right is simply right and wrong, wrong, and that a man who knows the difference between right and wrong must be considered as responsible for his actions, no matter what modifying circumstances or mental conditions may enter into the problem of the decision as to his responsibility. If the clergyman would but realise how difficult in any individual case must be such a decision, and how much must be known with regard to the previous character of the individual, then a great beginning for the modification of the present over-severe modes of thought will have been made.

From a theoretic standpoint, it would not be easy to state all that the physician considers necessary to enable him to make his decision as to individual responsibility. Perhaps, however, the consideration of a series of cases that have attracted widespread attention, and which have been most carefully investigated in all their circumstances, may present the methods of responsible determination better than any set of rules. Three presidents of the United States have been murdered within forty years. The murderers were native-born Americans. In none of the three cases was there any adequate motive for the commission of the crime. The assassin in President Lincoln's case might, it is true, be presumed to have a sufficient political motive, but no entirely sane man could have thought for a moment that any good would be accomplished at that time for the South by the removal of Lincoln. A man of known erratic tendency, with the craving for theatrical effects deeply ingrained in his nature, with a personal history not entirely free from even more serious manifestations of mental disequilibration, and with a family history of more than suspicious character as regards the mental qualities of his ancestors, committed the crime. He met his death at the hands of pursuing soldiers. Such was the temper of the time, that had he been captured alive he would surely have suffered the formal legal death penalty. Even as it was, public sentiment clamoured for legal victims and unfortunately they were found.

It seems clear, beyond all doubt, that in this case complete responsibility for his action was not present in the assassin himself. The courts decided later that there had been a conspiracy, but there has always been the feeling that justice was misled by over-zeal to find scapegoats for injured public sentiment. There is no doubt that it is an extremely difficult matter to say what shall be done to the assassin in such a case. The unfortunate result is as much an accident as the fatal consequences of any other perverted natural force. An earthquake may kill its thousands and the inevitable must simply be accepted. Society may protect itself from the further manifestations of such perverse individuals by confining the unfortunate murderer for life, but capital punishment, in the sense of a sanction for broken law, can scarcely be considered to have a place in the given conditions.

With regard to the murderer of our second assassinated President we had the farce of a long-drawn-out public trial of a man who was evidently not in his right senses. Once more a victim had to be found to satisfy injured public feeling. Guiteau was condemned to death and suffered the death penalty. Any one who reads the proceedings of the trial and who realises the significance of the motive that Guiteau himself gave for his act, will appreciate that the court had to do with an irresponsible doer of a material but not a moral wrong. There were many signs of mental disequilibration in Guiteau's previous career. It is on these eventually that the expert in mental diseases must depend in order to enable him to obtain a proper estimate of the extent of the mental disturbance in any given individual. It may seem that many real criminals can be defended on this same principle of finding an inadequate motive for their crimes. There are, however, certain signs of irrationality not difficult to detect if the previous life of the individual be carefully scrutinised and these must form the ultimate criterion as to criminal responsibility.

With regard to the third murderer of a President the case is clear. He was an ignorant, somewhat conceited individual, but he presented none of the signs of true mental disequilibration that can ordinarily be depended on to indicate such a disturbance of the physical basis of mind as impairs responsibility. He was not entirely without a motive, which in the mind of a brooding, conceited individual, might seem to be adequate for the commission of the crime. His sentence of the death penalty was then in accord with the judgment of the best mental experts. How society shall protect itself, and especially its high officials, against such notoriety seekers is hard to say.

The consideration of these cases gives a clear idea of how a physician endeavours to fill up gaps in his knowledge of the character of the man, his heredity and environment, as well as his previous actions at various times in life when under the stress of emotion. It may be considered that such a weighing of circumstances will serve to excuse many genuine criminals who eminently deserve to be punished. This is, however, the assumption of the older generation who considered that if a man did a material wrong he must be punished for it. It is a heritage of the day, when even accidental killing was considered to demand some punishment. At the present time the tendency is rather to consider only the moral wrong, that is, to calculate responsibility only for such actions as are committed with due deliberation, intention, and the knowledge of right and wrong as well as the freedom to perform the action. The old English legal opinion which declared a man responsible if he knows that what he is doing is wrong has now given way in most judicial proceedings to the principle that the man must not only know that he is doing wrong, but that he must also realise that he is free not to do that which he knows to be wrong. That is to say, if he feels himself compelled to the commission of crime, there is surely an impairment of responsibility. Such impulses to do wrong without adequate motive occur not infrequently among those whose mental condition is not perfectly normal, and this must always be taken into consideration in the ultimate decision as to their responsibility for their action.

JAMES J. WALSH.

SUICIDES

It is a very difficult problem at times to explain just how a suicide is due to mental alienation in a person whose intellectual powers appeared previously unimpaired, yet in most of the cases a knowledge of all the circumstances and of the individual himself would lead inevitably to this more charitable view. Most suicides are persons that have been recognised as paranoiacs and likely to do queer things for a long time beforehand. Indeed, some of the melancholic qualities on which the unfortunate impulse to self-murder depends are likely to have exhibited themselves in former generations. Not long since it was argued that the regular occurrence of a certain number of suicides every year—varying in various places, always on the increase, but evidently showing a definite relationship to certain local conditions —demonstrate that the human will is not free, since from a set of statistics one can foretell about how many cases of suicide would take place in a given city during the next year. As a matter of fact, suicides are not in possession of free will as a rule, but are the victims of circumstances and are unable to resist external influences.

The most important feature of suicide in recent years is the constant increase in the number, the increase affecting disproportionately young adults. This increase in the number of suicides is no illusion; it is not due to more careful statistics. It is true that in recent years, that is to say during the last quarter of a century particularly, the unsparing investigation by the authorities of all cases of suspicious death, and their report by sensational newspapers, has added somewhat to the apparent number of suicides. Families were accustomed to announce accidental death and have their story unquestioned, in a certain number of cases, where now there is no hope of concealment because of the unfortunate publicity that has crept into life. This increase, however, would account for only a small additional number of suicides, while the actual figures have more than trebled in the last thirty years.

This increase has come especially in the large cities. According to the report of the New York Board of Health, there were 1,308 suicides in New York City during the decade from 1870 to 1880. During the decade from 1890 to 1900 there were 3,944 suicides. This increase is much more than the corresponding increase in population. During the first decade mentioned there were 124 suicides per million of population. During the last decade this had risen to 196 suicides per million. The increase is nearly 60 per centum. The increase is variously distributed over the different ages. While every five years from twenty upwards shows a percentage of increase in the number of suicides committed, somewhat less than the percentage of increase for all ages, the five years between fifteen and twenty shows an increase of 106 per centum. In a word the deaths of adolescents from suicide have more than doubled in the last thirty years.

Towards the end of the last decade of the nineteenth century there was for a time a cessation of the continuous increase. This occurred during the years 1898 and 1899. Apparently it was due to the fact that the occupation of the country with other interests, the war and its problems, and the fact that an era of prosperity made material conditions better, and thus gave less occasion for depression of spirits. During the years since 1900, however, the increase has not only reasserted itself, but has more than made up for the period during which suicides were less frequent. The increase during the last four years is more than was noted during the six years from 1891 to 1897.

The same increase has been noted in European cities. The higher the scale of civilisation in a city, or at least the greater the material progress and the more strenuous the life, the higher the death rate. In Dresden, for instance, the rate is 51 suicides per 100,000 every year. In Paris it is 42, in Berlin it is 36; while in Lisbon and Madrid it is lowest, being only respectively two and three per 100,000 per year. While suicides are more common among men than women in all countries, this is not true for certain ages. Between the ages of fifteen and twenty-five the suicides of women are more numerous than those of men. The suicides of women are increasing faster than those of men. Fifty years ago the proportion was five to one. Twenty years ago it had fallen from three to one. Now it is less than two and a half to one. The saddest feature of the suicide situation is the increasing number of the children who commit suicide.

Almost needless to say, children's suicides are without any serious motives and are usually due to an attack of pique because of a slight from a playmate, a reprimand at home, a rebuke from a teacher, envy of the success of a companion, disappointment over a passing love affair, sometimes a peculiar attachment in the case of weak and morbid individuals, the manifestations of which are resented by its object, or are forbidden by parents and guardians. These unfortunate accidents have become so common now that special care must be taken with regard to children of neurotic heredity. When in previous generations there have been the manifestations of lack of mental equilibrium, then children's mutterings with regard to possible recourse to suicide should be the signal for the exercise of close surveillance. As far as possible such children should be kept from the strenuous competition at school in modern life.

As has been well pointed out there is no doubt that the power of suggestion and example has much to do with the increase of suicide. Dymond, an authority in the matter, says, "The power of the example of the suicide is much greater than has been thought. Every act of suicide tacitly conveys the sanction of one more judgment in its favour. Frequency of repetition diminishes the sensation of abhorrence and makes succeeding sufferers, even of less degree, resort to it with less reluctance."

Our modern newspapers, by supplying all the details of every suicide that occurs, especially if it presents any criminally interesting features or morbidly sentimental accessories, familiarise the mind, particularly of the impressionable young, with the idea of suicide. When troubles come lack of experience in life makes the youthful mind forecast a future of hopeless suffering. Love episodes are responsible for most of the suicides in the young, while sickness and physical ills are the causes in the old. In a certain number of cases, however, domestic quarrels, and especially the infliction of punishment on the young at an age when they are beginning to feel their independence and their right to be delivered from what they are prone to consider restriction, are apt to be followed in the morbidly unstable by thoughts of suicide.

The important practical question is the prevention of the fulfilment of the morbid impulse during these impressionable years. Many a young person has been saved from suicide at this time to realise the enormity of the act and to live without any further temptation to its commission for a long lifetime. As a rule the motive for the act is so trivial and often so insensate that it is not difficult to make patients (because patients they truly are) see the folly of their irrational impulse.

In order to forestall the putting into action of their impulse it is important that those who are close to the patient should have some realisation of the possibility of its occurrence. There are usually some signs beforehand of the possibility of the crime. Many of these early suicides have distinct tendencies to and stigmata of hebephrenic melancholia. The best known symptoms of this condition are those described by Dr. Peterson, the present president of State Commission of Lunacy of New York in his book on mental diseases. The symptoms noted are extraordinarily rapid and paradoxical changes of disposition. Depressed ideas intrude themselves in the midst of boisterous gaiety, and untimely jocularity in the deepest depression, or at solemn moments. Then there is the paradoxical facial expression, the so-called paramimia, that is, a look of joy and pleasure when really mental depression is present, or a look of depression when joyful sentiments are being expressed. The existence of such rather noticeable peculiarities may lead to the suspicion of mental disequilibration in young people.

The most important warning may well be the occurrence of suicide in any other member of the family for several generations before. The tendency of suicide to repeat itself in families is now well known and recognised. During the year 1901 in New York City, in one case other members of the immediate family had committed suicide in six instances. The subject has taken on additional interest because of the suicide of a well-known gambler who was the fourth of his family in two generations to take his own life. In another case, reported within the last five years, the suicide was the last of a family of nine children, every one of whom had committed suicide. There is the record in the German army of four generations of a noble family, the eldest son of which committed suicide during the 5 years from 50 to 55.

In these cases the tendency to suicide is not directly inherited, but there is a mental weakness that makes the individual incapable of withstanding the sufferings life may entail. In the later members of the family there is also a suggestibility that the frequent contemplation of the idea of suicide finally leads to the putting off of the natural abhorrence at the thought of its commission. In such families, therefore, it is particularly important to warn medical attendants and members of the family of the possibilities of unfortunate acts so as to prevent if possible the execution of any impulse to self-murder.

JAMES J. WALSH.

VENEREAL DISEASES AND MARRIAGE

Syphilis is a disease that is contagious, inoculable, and transmissible by heredity. It may be acquired innocently, and it is so acquired in about 4 per centum of cases according to good authority, but the other 96 per centum is venereal. The disease attacks any part of the body within and without from the soles of the feet to the hair and finger-nails. The first evidence, where the case is not hereditary, is a hardened sore called a chancre; next the lymphatic glands swell, and many forms of skin-eruption occur; then follows a chronic inflammation of the cellulo-vascular tissues and the bones, and small tumours, called gummata, may develop in almost any part of the body. The disease may vary from a light attack to malignancy. There are periods in the course of the disease.

1. The period of primary incubation, or the time from infection to the appearance of the chancre. This is commonly three weeks.

2. The primary stage: the chancre forms and the neighbouring glands are affected. This stage lasts from three to ten days.

3. The secondary incubation, or the time between the appearance of the chancre and the development of what are called the secondary symptoms,—usually about six months.

4. The secondary stage. Here occur fever, anaemia, neuralgic pains, and the eruptions on the skin and the mucous membranes. This period lasts from twelve to eighteen months in the majority of cases.

5. The intermediate period. During this time there may be no symptom, or slighter recurrence of the secondary symptoms. This period lasts from two to four years. It may end in recovery of health or be followed by tertiary symptoms.

6. The tertiary stage. In this period gummata form, or there may be diffuse infiltration of various parts of the body, chronic inflammation and ulceration of the bones, skin and other tissues, nervous diseases, and so on.

259

The tertiary stage commonly begins from three to four years after the primary infection.

The three chief divisions, which are apt to blend one into the other, are the primary, secondary, and tertiary periods.

The affections of the secondary stage are often severe. There may be fever associated with weakness, headache, general malaise and pain, and this may be marked or rather light. In this stage iritis is liable to occur, and if it is not properly diagnosed and treated it will result in blindness.

The lesions of the tertiary stage may cause great destruction of tissues and very grave consequences. Cerebral syphilis, if unchecked, will inevitably cause paralysis or paresis. There may be loss of speech, epilepsy, coma, paralyses, apoplectic hemiplegia, and so on. The pain is harassing and often it amounts to great anguish. Whatever part of the brain substance is destroyed will not be restored.

In syphilitic affections of the spinal cord, if the inflammation is acute death ensues in a few days or weeks. Tabes dorsalis, or locomotor ataxia, is caused in about 93 per centum of cases of this disease by syphilis, and it is an incurable and dreadful malady.

If there is neuritis from the virus it becomes intense and causes muscular contractions, paresis, and paralysis. The optic, auditory, and olfactory nerves may be attacked and destroyed. The nose also may be destroyed and it commonly caves in. The bones of the face are frequently attacked in the tertiary stage and they rot away. The tibia is diseased more frequently than the other long bones.

The heart is rarely injured, but when it is, the prognosis usually is bad. In a large number of cases death is sudden and unexpected. If the arteries are involved the prognosis again is bad, because the symptoms here do not show until too late for effective treatment When the liver is the seat of gummata these may be cured in the early stage, but in the later stage the prognosis is unfavourable. Some forms of renal syphilis are remediable, but others are not, especially the interstitial kind.

Syphilis is transmitted to a child congenitally, not as a tendency or predisposition, but as an active contagion. It may come from the father, the mother, from both parents, or by direct infection.

The transmission from the father is the most frequent. The spermatozoa carry the infection to the maternal ovum. Down to the end of the secondary stage, and half through the intermediate period between the secondary and tertiary stages of syphilis, a father or mother may beget a child that will be infected with hereditary syphilis, a shivering, blasted, rotten little wretch for whom a quick death is the greatest imaginable blessing, and it usually gets this blessing. In the acute stage of a virulent syphilis the disease is most likely to be transmitted; but sometimes, though rarely, a father that has been free from all symptoms of syphilis for many years may beget a child that is born with a virulent hereditary form of the disease.

Infection by the mother is more certain and more harmful than that from the father, because the intrauterine life of the child is poisoned throughout its course. Two-thirds of the cases of hereditary syphilis die either by abortion, or if they live to term they die shortly after delivery. If the mother is infected during the first eight months of pregnancy the child will nearly always be syphilitic, but if she is infected after the eighth month the child may escape.

If at the moment of conception both parents have the disease the child will almost certainly take it, and this infection will cause its death. In a series observed by Fournier, 28 per centum of the cases caused by paternal infections died and 37 per centum showed the luetic taint; in the cases caused by maternal infection 60 per centum died, and 84 per centum had syphilitic lesions; in the mixed heredity, that is when both the father and mother were luetic, 68.5 per centum died and 92 per centum were born syphilitic. When a child is first infected at delivery the case is not technically classed as hereditary syphilis.

During the first year after the father or mother has taken syphilis the probability of infecting the child is the greatest. In the third year the liability of infecting the child is lessened, but present. In a series of 562 cases of hereditary syphilis observed by Fournier, 60 children, over 10 per centum, were infected more than six years after the primary parental infection. Carefully observed cases have been exceptionally found where infection of the child has occurred in the fifteenth and even the twentieth year after the original parental lesion. Fournier reports the case of one woman that had nineteen consecutive stillbirths from syphilis.

Mild parental syphilis may transmit hereditarily the most malignant type of the disease, and very virulent parental infection may result in a comparatively mild infection of the child, if any infection by syphilis may be called mild. That the parent shows no symptoms from an old infection is no proof that he or she is cured, or that the child may not be infected.

With proper treatment of the mother the infantile mortality in hereditary syphilis is reduced from 59 per centum to 3 per centum, and the children that are born living are not unfrequently free from syphilis. When a woman is infected at the conception of her child miscarriage takes place before the child is viable, from the first to the sixth month; later other miscarriages occur; later still, living but syphilitic children are born, of whom one-fourth die within the first six months; finally she may have children that are not infected.

If a syphilitic man has been properly treated he may, after four years, beget healthy children, and he commonly does, but he may be the father of syphilitic children. Syphilitic women properly treated may, after about six years from infection, bring forth healthy children, and they commonly do, but not always.

There is a wide diversity of opinion among the best authorities concerning the curability of syphilis. Gowers (Syphilis and the Nervous System. 1892) says: "There is no evidence that the disease ever is or ever has been cured, the word 'disease' being here used to designate that which causes the various manifestations of the malady." He means there is no absolute proof that a person who has once been infected is ever so fully cured that he may marry without danger of transmitting the disease.

Fournier requires, as the minimum time, four years of methodical treatment before he deems the patient safe, but even this arbitrary fixing of the number of years is not warranted by experience. Many physicians hold that in the tertiary stage the disease is not transmissible, but that statement is not true. Commonly it is, sometimes it is not. After all symptoms have disappeared the disease has been transmitted.

In short, a person that wittingly marries any one who has had syphilis at any time is a fool; and if one of the contracting parties has had syphilis within the four years preceding the marriage the marriage is criminal, even if the syphilis has been carefully and skilfully treated by a physician.

Gonorrhoea is always a dangerous disease. In the male, beside the acute lesions, it can cause chronic or fatal inflammations along the various parts of the genito-urinary tract or in different organs of the body. When the disease becomes chronic it lasts indefinitely. It may then cause cystitis, or so affect the kidneys as to bring about very grave results; it may get into the circulation and induce gonorrhoeal rheumatism of the joints, especially of the knee joint, and result in a partly or completely stiffened joint. The heart may be affected and endocarditis ensue; there may be meningitis or inflammation of the cerebral membranes; the eye may be infected, and unless it is skilfully treated blindness will follow. Strictures of the male urethra from chronic gonorrhoeal inflammation often require major surgical operations for relief.

The disease in women has most of these complications, and other grave peculiar phases. All prostitutes have acute or chronic gonorrhoea, and 12 per centum, probably more, of reputable women are infected; and the suffering caused is very great. The gonococcus remains virulent for two or three years at the least in a man's chronic gleet, and if he marries he infects his wife. Should her womb be infected she is seldom completely cured. If the Fallopian tubes are involved, and this happens frequently, they suppurate, and often they must be removed by coeliotomy. The woman suffers for a long time when the tubes are attacked by the disease, and she becomes sterile ordinarily.

When a child is born to a woman that has gonorrhoea its eyes are infected at delivery, and if it is not very skilfully treated it will surely lose its sight. Because of this danger, in maternity hospitals the eyes of all babies are treated at delivery as a precaution, and many physicians observe the same precaution in private practice.

When, therefore, a man has chronic gonorrhoea he should not marry until about four years after the last infection, and he should be carefully treated in the meantime. There is a popular opinion that gonorrhoea is a trifling disease, but the contrary is the truth: it is a grave disease, especially in women; and the person that carelessly infects another is certainly guilty of crime for which a long term in jail would be a light punishment.

AUSTIN ÓMALLEY.

SOCIAL DISEASES

There are certain affections not at all uncommon and as a rule producing rather serious effects upon the social body, of which, though their existence is well known to all, very little is said. It is certain that what is considered the more severe of these venereal diseases may be acquired quite innocently. Indeed, many thousands of cases of this affection, acquired innocently, have now been reported by medical men in this country alone. If the statistics of all the world were gathered together, there would probably be a hundred thousand cases of this dreadful affection, which have been acquired without any blame on the part of the sufferer. It has become the custom, especially in English speaking countries, to ignore the presence of these diseases, and this has led to a multiplication of opportunities for their spread to such a degree that now the condition of affairs, for those who know it best, is rather alarming. It is with the intention that a few definite ideas, given absolutely without exaggeration and without any striving after effect, may enlist clergymen, as well as physicians, in a crusade against these diseases, that the present chapter is written.

It has been said over and over again at medical society meetings that it is a very unfortunate thing that universities in these modern times are situated in large cities. The young man just freed from the restraints of home life, or of the seclusion of a college, is at once without any preliminary training, exposed to all the dangers, moral and physical, of large city life. Not only is this true, but he is even not properly warned of the dangers that lie so close to his path. Our prudery has gone so far that the very names of these affections are tabooed and above all must not be mentioned before the young. As to the awful evils that such diseases may cause, as to the lifelong suffering, even to mental degeneration and early death, that they may involve, not even a hint is considered to be proper. The consequence is that

265

young men expose themselves not infrequently to danger, absolutely unknowing the significance that such diseases have in recent years acquired in the minds of modern physicians, and it is usually not until a serious mistake has been made that the young man is brought in contact with the physician who may be frank in pointing out evils utterly unknown before.

This state of affairs has come to be considered as so irrational in many foreign universities that now a special course of lectures is given every year on the significance of what may be called the social diseases. The students are told very frankly what the possibilities of danger for them in certain excesses may be, so that at least the young man can not say "I knew nothing about it," when the risk becomes an actual reality of danger. At the University of Berlin the first course of such lectures was established, and the interest aroused and the results obtained were such as to make other universities consider the advisability of such lectures for their students. Even here in prim and prude America, one or two of the great universities have come to the realisation that the physical well being of their students is committed to their care, as well as their intellectual development, and at least a beginning of that precious wisdom that comes from the fear of the physical evils of sin has been acquired because of opportunities provided by the faculty.

It is well admitted now by all that ignorance is not innocence and that knowledge of the consequences of social diseases is likely to be a very important factor in preventing young men from taking risks that would otherwise be considered very slight, perhaps. As a matter of fact, nothing can be more helpful from the ethical standpoint than this knowledge of how closely may follow the wages of sin, which is death. It is for this reason that clergymen would seem to owe it, as a duty to themselves and their position in social life, to acquire a certain knowledge of these affections. A very great change has come over the attitude of the medical profession towards the so-called venereal diseases in recent years. A quarter of a century ago they were considered to be not very serious after all, and indeed in some cases to be no more serious than a cold, a mere passing incident in life. Now it is well recognised that almost never do they leave their victim in the state of health in which he was before, and that unfortunately the deterioration of tissues which has taken place is likely to be enduring. Even many years afterwards there may be serious complications involving health or even life.

For instance, it is now very generally conceded that paresis, or what is sometimes called general paralysis of the insane,—a progressive mental and nervous disease, which invariably ends fatally in from three to seven years,—is always due to one of the so-called social or venereal diseases. How important this affection has become in modern life can be best appreciated from the fact that in Europe nearly one in four of those who die in the insane asylums are sufferers from paresis. In this country the disease is not so frequent, the proportion being less than one in five or even one in six. The disease is becoming more and more common, however, as large city life becomes more prominent, and as the possibility of infection with social diseases is more widespread.

Paresis is what is sometimes called softening of the brain, and it attacks by preference men under thirty-five. The first symptoms of it as a rule are not alarming. A young man's disposition changes, so that an individual heretofore rather stingy becomes extravagant, while occasionally a prodigal becomes very saving and considers that he has already a large sum of money to his credit. The most prominent feature of the early stage of the disease is the occurrence of delusions of grandeur, that is to say, the patients get the idea that they are important personages, or that they have fallen heirs to a large sum of money, or that they have been appointed to high salaried positions. As a consequence of these delusions, they may make expensive presents to their friends. Occasionally there are other changes in disposition. A young man, for instance, who has been of genial character becomes morose and hard to live with. The opposite change to greater liveliness of disposition is not unknown, but is more infrequent. Sometimes there are marked excesses, high living, luxurious habits, and the like, before the existence of disease is recognised.

The mental stigmata of the disease at the beginning are not alarming at all. There are slight lapses of memory. A man who has hitherto been known as an accurate mathematician, makes frequent mistakes in adding or multiplying. The physical signs are even slighter. In using long words, syllables are omitted from them. A favourite method of testing the speech of a person suspected of beginning paresis is to ask for the pronunciation of a word like Constantinople. Usually a syllable will be elided, and the reply will be "Constanople," or something similar. There is a slight tremulousness of the hands and usually a rather easily marked tremor of the lips, especially when the tongue is protruded. Often in the very earliest stage of the disease, there are changes in the pupils. They may be unequal, or may fail entirely to react to light.

When these signs are positive, that is, when there is a change in disposition and then the physical stigmata that we have gone over appear, the diagnosis of the disease is almost certain. The physician is able to say, with considerable assurance, that the young, strong, healthy-looking patient, who has often had to be tempted to come to see the doctor by some specious reason, because he does not consider himself that he has anything wrong with him, will have to be confined in an asylum within a year, or at most, two, and will die in a state of dementia within five years. This, of course, is an awful picture. This is the course of the disease in nearly 20 per centum of the inmates of our asylums. Almost without exception there is a history of syphilis in these cases, and the medical world is now persuaded that this is the most important factor in the production of paresis.

Another nervous disease, corresponding in some of its features to paresis and indeed sometimes spoken of as a spinal form of paresis, is locomotor ataxia. This affection begins usually with loss of sensation in the soles of the feet so that the patient thinks that he is walking on carpet all the time. Before this there may have been some disturbance of vision. The pupils may fail to react to light. Occasionally the first symptoms are motor, that is, the man notices that he is not able to walk as readily as before. He staggers easily. If he tries to turn round while walking he is apt to lose his balance. If he tries to walk in the dark, he is almost sure to have so great a sense of insecurity that he dare not go far from the wall. Occasionally the first sign is a sinking of the limbs on the way down stairs. In certain very sad cases, the first and only symptom is a failure of sight which goes on progressively, until the optic nerve is completely destroyed and sight forever rendered impossible.

All these symptoms are traceable directly to certain changes which have been noted in the spinal cord. These changes are due to disturbance of the blood and lymph supply of the nervous tissue. Once the changes have taken place, there is no hope of the patient ever recovering the normal use of his limbs. Not infrequently he becomes bedridden and can not walk at all because he is not able to steady himself. He may not suffer in his general health, however, to any serious extent, and may live on for twenty years, though usually his resistive vitality is lowered and he is carried off by some intercurrent disease.

At times locomotor ataxia begins with very severe pain seizures, known as crises. These may occur in the legs or arms or in the stomach or sometimes in other organs. Occasionally they are the first symptoms of the disease that are noticed, and they may continue for months or even years before other symptoms manifest themselves. This sometimes makes it difficult to recognise the disease for what it really is. The pains are usually most excruciating, are tearing or boring in character, and are sometimes described by the patient as being similar to the sensation that would be felt if a red hot iron were forced into them, or if a knife were inserted and then twisted round. Hence the descriptive name which has been applied to them of "lightning pains" which describes the suddenness of their onset and the intensity of their character. Most of the ordinary anodyne or pain-killing medicines fall to influence them, and the patient is one of the most pitiable of objects while they last.

It is now conceded on all sides that at least 75 per centum of the cases of tabes are directly due to syphilis. Indeed this affection and paresis are sometimes spoken of as parasyphilitic affections. Unfortunately the ordinary treatment for syphilitic manifestations does not affect them in the least. So far as we know at the present time, there is nothing that will hinder the course or prevent the progress or alleviate the symptoms or have any curative action on either of these dreadful diseases. They are much more common in Europe than they are in this country, but have been seen here with quite sufficient frequency in recent years to make physicians, at least, realise the necessity for having young men appreciate the dangers they invite in thoughtlessly yielding to the temptations of great city life.

There are other affections which can be traced directly to the social diseases. One of the most important of these consists of certain brain tumours which may even cause death if not properly treated. These syphilitic brain tumours frequently cause paralysis and may lead to permanent changes in the nervous system with consequent loss of motor power. Whenever the symptoms of brain tumour occur, careful inquiry is made as to the previous existence of syphilis in the case, in order to determine, if possible, if this is the morbid agent at work. If there is a history of syphilis it is usually said to be fortunate, for brain tumours due to syphilis may be made to disappear by the proper use of mercury and the iodides. If the treatment of the case is delayed, however, alterations in the nerve substance take place which can not be improved.

This disease affects especially the blood vessels and, as a consequence of the thickening of the coats of the arteries, blood may be shut off from certain portions of the brain entirely. This will, of course, produce symptoms of paralysis. Indeed, whenever paralytic symptoms manifest themselves under forty years of age, the physician's first thought is sure to be that there is syphilis in the case. This is not always true, for by heredity and very hard work occasionally arteries become so degenerate that they rupture before a patient has reached many years beyond forty, but the case is always suspicious. In this, as in the corresponding instance of brain tumour, treatment, if applied sufficiently early, may not only give relief of all the symptoms, but produce a complete cure. That is, at least the symptoms are relieved for the time, though there may be relapses. Usually these relapses are quite amenable to treatment, but sometimes they get beyond the control of the physician and death ensues. It is almost the rule where there have been serious nervous symptoms once, that recurrences of them must be feared, and they will eventually shorten the patient's life.

Syphilitic manifestations of serious character develop, however, not only in the nervous system, but also in certain of the important internal organs. The liver may become so much affected as to refuse to do its work. Solid tumours may develop in the stomach, or along the course of the intestines, resembling cancer so much that occasionally operations are performed for their removal. As a rule, however, these yield quite promptly to proper antisyphilitic treatment. Whenever an obscure intraabdominal tumour is present, accordingly, it has become the custom among physicians and surgeons not to make an absolute diagnosis nor to perform any serious operation until antisyphilitic treatment has been tried. The surprises of such treatment constitute a very interesting chapter in obscure diseases in medicine.

As we said at the beginning, it is perfectly possible to have contracted the disease innocently, and indeed, the first manifestations may be so mild as to fail to attract the patient's attention. In these cases there will be no history of syphilis, yet the test of antisyphilitic treatment will demonstrate that the disease has been present. Not a few physicians have died from these serious manifestations of syphilis after having contracted the disease through a cut on the finger or the prick of an infected needle in the ordinary course of their professional work. Some of these cases in young men prove to be especially malignant and fail to react to treatment, so that a fatal issue takes place within a few years.

On the other hand, in general it may be said that the disease is eminently curable, though it may require great care on the part of the patient and the avoidance of all excesses either of work or indulgence for the rest of life. It has often been noted that people who live in the midst of serious emotional strain are most likely to suffer from manifestations of syphilis in their nervous system. Hence it is that paresis and locomotor ataxia are comparatively quite common among actors, brokers, and financiers. They are also quite common among sea captains and military men who are exposed to severe hardships and have to assume weighty responsibilities. In such men the previous attack of syphilis has so weakened the nervous system that it degenerates under the strain placed upon it by the subsequent responsibilities. These diseases are very uncommon among clergymen and are less common in Ireland than in any other country in the world, which would serve to confirm the opinion that the venereal disease is a prominent factor in their causation.

We would not have the idea be assumed that syphilis is an incurable disease and is bound to be followed in all cases by the awful manifestations that we have described. There are many thousands of cases of syphilis that never have any of these serious manifestations at all. It is evident that some cases are completely cured and that no deleterious influence remains. On the other hand, it must not be forgotten that the presence of this disease in the tissues of either parent during the first five years of its course are almost sure to affect offspring born at this time. The children may suffer from the skin lesions of syphilis in their early life, may suffer from serious eye diseases a little later, and then eventually succumb to nervous and mental diseases resembling paresis and locomotor ataxia in early adult life. In fact it is this transmission of the disease that constitutes one of its saddest pictures, and the sins of the parents are indeed visited on the children.

Besides this severer type of social disease, there is what has been called sometimes a milder form. It consists only of a discharge with some fever, which is considered to last not more than a few weeks. As a matter of fact, however, the disease may continue to exist, though the symptoms become latent and the patient may infect others when he least suspects it. This form of disease gives rise to many sad complications in family life. Practically all the severe eye diseases of newly born children, the ophthalmia from which so many eyes are lost, is due to this disease. Special medical care is now taken of these cases, and the serious consequences are not so often seen as used to be the case. Within a score of years, however, about one-half of the inmates of blind asylums owed their loss of sight to this disease. At the present time there still remains a very notable proportion of persons blind from early childhood whose infirmity must be attributed to the sad consequences of the social disease.

Most of the sterility in families is due to the same cause. There is an unfortunate impression that usually the woman is responsible in these cases, and not a little sympathy is wasted on the man, because of the absence of children in the family. Almost invariably, however, the real cause of the family misfortune is to be traced to an infectious disease in the man contracted perhaps many years before, of whose presence he may be more or less unconscious, the symptoms have become so slight, but this has proved sufficient to infect the wife and bring about serious changes that preclude all possibility of the procreation of children.

These statements may seem exaggerated. On the contrary, they are rather understatements of actualities. No one who knows the real state of the case will fail to realise this. Physicians themselves have only come properly to appreciate the true state of affairs in the last twenty years. We need a coordination of all the forces that make for social amelioration in modern life to correct present false impressions.

JAMES J. WALSH.

DE IMPEDIMENTO MATRIMONII DIRIMENTE IMPOTENTIA

Hoc argumentum praecipue ad juris consultos ecclesiasticos et civiles pertinet; et quamvis differentia sit inter jurisdictionem judicis civilis et ecclesiastici tamen judicium utriusque quatenus necessario pendet ab existentia conditionum physicalium in medici consilio situm est. Obscuritas doctrinae et quidem gravis de hoc impedimento, libris moralistarum, medicorum et juris consultorum perlectis, invenitur; et quamvis, elapsis perpaucis annis, fere omnis liber tractans de scientia medicinali parva fide dignus, tamen multa ex editis physiologorum veterum tanquam vera a moralistis praesertim promulgantur. Hae difficultates per ignorantiam anatomiae et physiologiae genitalium non minuuntur. Ut auxilium, si quid sit, ad difficultates solvendas feram, species et gradus Impotentiae hie collegi tanquam medicus, eo modo ut conditio physica clarius cognoscatur.

In unoquoque Statuum Foederatorum Americae Septentrionalis impotentia ratio sufficiens divortium obtenendi est, in plurimis autem matrimonium irritum ab initio non reddit. Impotentia vel temporanea causa divortii esse potest si impotens intra spatium temporis rationabile remedium medicinale recuset. Sub lege civili Americana contrahens qui tempore matrimonii ineundi certior erat de impotentia consortis jus divortii petendi propter abnormalitatem istam amittit. Procrastinatio longa et inexplicabilis divortii petendi, et etiam inscitia culpabilis impotentiae consortis divortium impossibile coram judice civili reddunt.

Conditio haec etiam impedimentum dirimens matrimonii sub lege canonica Ecclesiae est. Si impotentia contractum matrimonii anteat et perpetua sit, matrimonii contractus solvitur ipso facto, quandocumque detegitur. Procrastinatio aut ignorantia culpabilis non excusant.

Jurgia oriuntur ex eo quod impotentia cum sterilitate saepius confunditur. Juris consulti civiles infrequenter hoc modo offendunt, medici autem et moralistae crebro in errore isto versantur. Juris consulti Americani et medici de impotentia doctrinam accipiunt librorum praesertim On Domestic Relations, auctore Irving Browne (Boston. 1890), A System of Legal Medicine, auctore Allen MacLean Hamilton (Neo-Eboraci, 1897), et A Manual of Medical Jurisprudence, auctore A. S. Taylor (Neo-Eboraci, 1897). Irving Browne (op. cit.) ait: "Ubi Impotentia adsit nullum habetur matrimonium validum. Impotentia autem incapacitatem prae se fert physiciam, non meram frigitatem, declinationem seu repugnantiam, neque etiam recusationem absolutam coitus sexualis. Neque sterilitas nec malformatio quae copulam non impediant, neque infirmatio quaecumque sanabilis incapacitatem gignunt. Impotentiam tempore ineundi matrimonii exstitisse necesse est." Eadem est doctrina Schouleri et Baldwinii.

White et Martin, medici, (Genito-Urinary and Venereal Diseases. Philadelphiae. 1897.) impotentiam ita definiunt: "Inabilitas actus sexualis perficiendi. Non necessario cum sterilitate consociatur, neque necesse est quod sterilis impotens sit." Et ita alii omnes.

Significatio vocis Impotentiae sub lege canonica deducitur, (1), ex dijudicationibus Pontificum Romanorum, aut (2), ex judiciis Congregationis Sancti Officii, tribunalis ad sententias hujus generis pronuntiandas instituti, aut, (3), ex legis interpretatione a moralistis scientia praeditis.

Sixtus V, Pontifex Romanus, (Const. Cum frequenter, anno 1587) decrevit eunuchos impotentes esse sensu legis canonicae de Impotentia, nullum autem judicium papale totam questionem conficit. Congregatio etiam Sancti Officii in perpaucis casibus particularibus dijudicavit sed legem nullomodo distincte definiebat. Norma igitur a nobis sequenda ex interpretatione moralistarum est depromenda.

Lex non est decretum mere disciplinare: e natura ipsa contractus matrimonialis desumitur. Ballerini (Theol. Mor., vol. 6, p. 658) scribit matrimonium consistere "in mutua traditione potestatis ad copulam conjugalem." S. Thomas (Supplem. Sum, Theol., q. 58, a. 3) ait: "In matrimonio est contractus quidam, quo unus alteri obligatur ad debitum carnale solvendum: unde sicut in aliis contractibus non est conveniens obligatio si aliquis se obliget ad hoc quod non potest dare vel facere, ita non est conveniens matrimonii contractus, si fiat ab aliquo qui debitum carnale solvere non possit; et hoc impedimentum vocatur impotentia coeundi."

Antequam explicationem a moralistis pleniorem vocum "Impotentia coeundi" dabamus, attendendum accurate est ad definitionem matrimonii finum a S. Alphonso Liguorio (Theol. Mor., lib. vi., n. 882) datam. "Fines," inquit, "intrinseci essentiales [sc. matrimonii] sunt duo: traditio mutua cum obligatione reddendi debitum, et vinculum indissolubile. Fines intrinseci accidentales pariter sunt duo: procreatio prolis et remedium concupiscentiae. Fines autem accidentales extrinseci plurimi esse possunt, ut pax concilianda, voluptas captanda, etc. His positis, certum est quod si quis excluderet duos fines intrinsecos accidentales, non solum valide, sed etiam licite posset quandoque contrahere; prout si esset senex et nuberet sine spe procreandi prolem, nec intenderet remedium concupiscentiae; sufficit enim ut salventur fines substantiales, ut supra."

Haec sententia S. Alphonsi magni momenti est, et in ea solutio multarum difficultatum inveniri potest. Dicit hic (1) fines intrinsecos essentiales matrimonii esse traditionem mutuam cum obligatione reddendi debitum, et vinculum indissolubile, atque illis demptis nullum matrimonium; (2) procreationem autem prolis et remedium concupiscentiae abesse posse, et tamen matrimonium esse validum si duo fines essentiales adsint.

Sanctus hoc loco infert, ut patet e contextu alibi (e. g., lib. vi., n. 1095, res. 2), traditionem mutuam potestatis ad copulam carnalem necessario potentiam coeundi supponere, potentiam autem generandi non esse necessariam nec remedium concupiscentiae. In libro vi., n. 1096, ait: "Impotentia est illa propter quam conjuges non possunt copulam habere per se aptam ad generationem; unde sicut validum est matrimonium inter eos qui possunt copulari, esto per accidens nequeunt generare, puta quia steriles aut senes, vel quia femina semen non retinet, ita nullum est matrimonium inter eos qui nequeunt consummare eo actu, quo ex se esset possibilis generatio."

Distinctio haec inter potentiam coeundi et potentiam generandi a moralistis omnibus datur; illa autem data, plurimi distinctionem obliviscuntur et sterilitatem simplicem cum impotentia confundunt.

A. Konings, C.SS.R., (Theol. Mor., ed. 7, vol. 2, p. 276) haec habet: Impotentia est "incapacitas ad copulam carnalem, per se aptam ad generationem." In n. 1619, § 5, ait: "Non est confundenda impotentia coeundi cum impotentia generandi. Hinc steriles et senes qui matrimonium consummare valent, valide contrahunt, item mulieres quae possunt semen excipere, etsi illud non retineant." Hanc doctrinam S. Alphonso refert (Theol. Mor., lib. 4, n. 1095, ed. Mech. 1845), et paragraphum hoc modo complet: "Non tamen carentes utero vel vagina." Hoc est, tenet mulierem utero et vagina carentem impotentem esse. Unusquisque carentiam vaginae impotentiam esse admittit; mulier autem sine utero semen excipiendi capax est, concupiscentiam quoque maris satiare potest. Sterilis tantum est. Potentiam etiam habet coeundi, semen excipere potest et retinere, concupiscentiam quoque satiare potest, etiamsi uterus, ovaria et tubi Fallopiani absint. Praeterea, illi duo fines intrinseci essentiales matrimonii existunt.

Augustinus Lehmkuhl, S.J., (Theol. Mor.), alius illustris discipulus S. Alphonsi est Impotentiam definit: "Defectus propter quem conjuges non possunt copulam habere per se aptam ad generationem." Alibi (American

Ecclesiastical Review, vol. 28, n. 3), de impotentia excisioneque ovariorum scribens, ait: "Puto, questionem propositam, utrum excisio ovariorum vel uteri constituat impedimentum dirimens necne, theoretice nondum esse plane solutam." Existimat autem questionem practice solutam esse judiciis Congregationis S. Officii, d. 3 Februarii, 1887, et d. 30 Julii, 1890, editis, matrimonium mulieris ovariis carentis et mulieris utero et ovariis carentis, permittentibus. Etiamsi haec judicia non edarentur tanquam leges formaliter generales, Lehmkuhl opinatur in casibus ejusdem generis aptari posse. Re quidem vera illa doctrina sequi potest practice et theoretice; nulla enim est quaestio seria de impotentia in muliere carente ovariis.

Joseph Antonelli tamen (De Conceptu Impotentiae et Sterilitatis relate ad Matrimonium, Romae, 1900) tenet carentiam ovariorum esse impotentiam sub lege; et Casacca (Amer. Eccl. Rev., vol. xxvii, n. 6, et alibi) eamdem opinionem sequitur. E contra, Marc (Inst. Mor. Alphon.) docet carentiam ovariorum uterique non esse impotentiam.

Joseph Hild (Amer. Ecc. Rev. vol. xxviii., n. 6) optime vindicat opinionem, nempe, carentiam ovariorum non esse impotentiam, et in corpore tractatus citat definitiones impotentiae a moralistis egregiis prolatas.

Schmalzgrueber (Theol. Mor., lib. iv., tit 15, n. 31) dicit: "Sola impotentia ad copulam dirimit matrimonium, non vero impotentia ad generationem."

Coninck (De Sacr., vol. ii., d. 31, dub. 7, n. 86) ita habet: "Steriles ... si aliter potentes sint ad usum matrimonii, valide contrahunt; quia nec generatio nec potestas generandi est de essentia matrimonii."

Mastrius (Dis. de Matr., q. v., n. 114) ait: "Impotentia est inhabilitas perpetua ad consummandum matrimonium ... non est ex eo praecise quod alteruter conjugum aut uterque sint steriles, quia impotentia ad generandum seu ad prolificandum, dummodo adsit potentia ad copulam carnalem et seminationem, non est impedimentum dirimens, ut omnes passim concedunt cum Scoto ... et ubi est certa impossibilitas ad bonum prolis, tunc matrimonium est ibi in remedium, non in officium."

Vincentius de Justis (De Dispens. Matr. lib. ii, c. 17, nn. 1, 2, 3) scribit: "Impotentia ad matrimonium est duplex. Prima, quae sterilitas dicitur, efficit ut proles generari non possit, ex se tamen matrimonium nec impedit nec dirimit, ut docent Sanchez, Guttier, Coninck. ... Ratio est, quia nec generatio, nec generandi potestas sunt de essentia matrimonii."

S. Thomas (Supplem, q. 58, a. 1) in articulo de Impotentia, quam Frigiditatem et Impotentiam Coeundi nuncupat, nihil de sterilitate scribit, nec de impotentia generandi tanquam quid impotentiae coeundi oppositum.

In omnibus hisce definitionibus verba de se, ex se, per se, et alia similia, adhibentur de copula carnali qua copula. Amort (De Matr, q. 101) de his verbis loquens ait: "Impotentia est inhabilitas corporalis ad copulam carnalem de se ad generationem prolis idoneam.—Dicitur: de se; potest enim contingere per accidens, v. g., ob debilitatem spirituum seminalium in viro aut femina, vel ob indispositionem matricis in muliere, quod copula carnalis, etiam perfecta, hoc est, per effusionem seminis in vagina mulieris completa, non sit idonea ad generationem prolis." Loquuntur moralistae, ut dixi, de copula carnali quatenus copula est sine respectu ad possibilitatem generandi.

Hisce omnibus positis, rogamus:

(1), Quid sit impotentia sub lege in muliere?

(2), Estne mulier carens ovariis, utero vel tubis Fallopianis impotens?

(3), Quid sit impotentia sub hac lege in viro?

(4), Estne vir aspermatosus impotens, et quid de viris semen sterile habentibus?

I. Impotentia Mulieris. Mulieres steriles frequentius quam viris, viri autem impotentes frequentius quam mulieres sunt. Impotentia absoluta et perpetua raro in mulieribus, in viris crebro invenitur.

In fundo pelvis femineae septum est a latere in latus, rectum inter et vesicam urinariam, et in medio hujus partitionis uterus, qui piroformis est, quasi ad perpendiculum jacet et cervix sua in vaginam intrat.

A cornibus uteri, i.e., ab angulis superioribus, tubi Fallopiani procedunt ad libellam, et apud terminos tuborum ovarium est in utroque latere. Tubi aperti sunt prope ovaria, et non substantiae ovariorum continui. Si unum ovarium et tubus oppositi lateris demantur, vel si tubus iste occludatur, ovum ex ovario manente migrare per partem exteriorem uteri et foecundari potest.

Genitalia externa mulieris e labiis majoribus praecipue constant; intra et inter haec labia minora seu nymphae sunt. Intra labia minora ad summum versus clitoris est; infra hanc est meatus urinarius; infra meatum, orificium vaginae. Per imam partem orificii in virginibus extenditur pellicula tenuis quae hymen vocatur. Haec communiter in coitu primo rumpitur.

Tempore mensium praesertim ova egrediuntur ex ovariis in tubos Fallopianos et inde in uterum. Si ova non foecundentur per vaginam amittuntur. Foecundatio in tube Fallopiano fit.

In muliere impotentia temporanea aut perpetua oriri potest e causis sequentibus:

(1), propter hymenem imperforabilem aut cribriformem, aut septatum aut annularem;

(2), propter vaginam duplicem;

(3), propter vaginismum aut dolorem;

(4), propter uterum prolapsum aut productionem cervicis uteri;

(5), propter atresiam vaginae aut labia adhaerentia;

(6), propter orificium vaginae in loco abnormali;

(7), propter arctationem vaginae;

(8), propter tumores aut incrementum morbidum intra vel circa genitalia;

(9), propter Sadismum;

(10), propter Masochismum;

(11), propter Sodomiam gradus secundi seu defeminationem; Sodomiam gradus tertii; Sodomiam cum horrore; Urningismum; Androgyniam.

1. Aliquando vice hymenis normalis invenitur membrana densa seu cartilaginosa, aut membrana continua, aut cribro similis, aut tanquam septum, aut annularis, quae impediat intromissionem. Operatione simplici chirurgica conditio removetur.

2. Raro septum adest quod vaginam in duas partes dividit et impotentiae causa est. Chirurgus septum removere potest.

3. Vaginismus contractio spasmodica musculorum ad orificium vaginae est, et haec vaginam claudit. Frequenter inter neuroses ideopathicas includitur, scrutinium autem diligens locum inflammationis detegit qui origo est spasmi reflexi. Insolenter conditio ex hyperaesthesia murorum vaginalium inducitur.

Medicatio vaginismi examen expertum supponit et quandoque scrutinium endoscopicum vesicae urinariae. Fissurae in ano, endometritis chronica, urethritis granosa circa cervicem vesicae urinariae, causae principales sunt vaginismi, et istae causae sanabiles sunt. Inflammatio acuta vulvae, vaginae, uteri, tuborum aut ovariorum, carunculae urethrales, urethritis, fissurae cervicis vesicae urinariae, haemorrhoides, fissurae recti, coccygodinia, ulcera uteri et amotio uteri vel ovariorum e loco debito, possunt tantum dolorem in coitu infligere ut mulier practice impotens fiat. Morbi autem isti fere omnes medicabiles sunt, sed tamen aliqui omnino pervicaces sunt. In vaginismo hysterico et in aliis casibus insanabilibus intromissio fieri potest ope chloroformi ad evitandum divortium.

4. Quando uterus prolabitur vel cervix producitur ita ut copula impossibilis sit, chirurgus mederi potest.

5. Atresia vaginae est occlusio vaginae in longum perfecta vel imperfecta. Nullum orificium invenitur. Congenita esse potest, et tunc plerumque desunt omnia organa generativa. Atresia etiam consequitur vulnera et inflammationes morborum, ut diphtheritis et scarlatina.

Ubi atresia per totam vaginam extenditur nullum datur medicamentum, et impotentia absoluta et perpetua adest. Labia adhaerentia separari possunt.

6. Aliquando sed perraro vagina in rectum aperit. Possibilitas removendi impotentiam e loco orificii vaginae pendet.

7. Arctatio vaginae oritur ex causis atresiae, et remotio conditionis quum impotentia inducat impossibilis esse potest.

8. Tumores pudendi, vaginae et recti, hypertrophia labiorum et clitoridis, et elephantiasis labiorum copulam impedire possunt. Alii tumores et hypertrophiae removeri possunt, alii autem insanabiles sunt.

9. Dantur perversitates sexuales quae viros impotentes reddunt, et haec aliquando in mulieribus inveniuntur. Tales sunt Sadismus, Masochismus, et

gradus Sodomiae praeter primum. Isti morbi animi causae sunt impotentiae in muliere propter aversionem ejus virorum etiamsi physice potens sit. Vix in matrimonium iniit talis mulier, et igitur perversitates istae parvi momenti relate ad mulieres, relate autem ad viros magni momenti sunt. De his quid infra dicendum erit.

10. Senectus nunquam reddit mulierem impotentem, virum autem reddit. Etiamsi nihil ad impotentiam pertineat, hic juvat dicere in locis temperatis terrae parturitionem maxima ex parte desinere anno a natu quadragesimo-quinto. Cessare potest anno vicesimo-octavo, et perstare post annum quinquagesimum. J. Whitridge Williams (Obstetrics. Neo-Ebor. 1903), casum citat mulieris quae anno sexagesimo-tertio aetatis puerum vicesimum-secundum peperit et postea menses aderant. Parturitio aliquando decem vel duodecim annos post ultimos menses evenit.

Nunc, estne mulier ovariis carens impotens? Nullo modo: sterilis tantum est. Nam (1) Congregatio S. Officii in matrimonium duas mulieres ovariis orbatas inire permisit. (2) Talis mulier capax est copulae carnalis aeque ac mulier habens ovaria, et moralistae omnes concedunt nil amplius requirendum ut matrimonium validum fiat. (3) Si talis mulier impotens sit omnis mulier insanabiliter sterilis impotens esset, et discrimen a moralistis prolatum impotentiam inter et sterilitatem nugatorium esset et puerile. Mulier in qua tubi Fallopiani occludantur sterilis est perpetuo; idem dicendum de muliere habente uterum infantilem, vel ovaria rudimentaria, vel ovaria morbida, et ita porro. Nemo autem tales tanquam impotentes unquam tenet.

Aliquando mulier ovariis orbata sensationem sexualem possidet, vulgo autem non possidet. In utroque casu tamen remedium idem concupiscentiae mari perstat, et hoc sufficit pro viro ut matrimonium validum sit. Huc accedit, relate tum ad mulierem tum ad virum, quod duo fines intrinseci essentiales matrimonii, et fines sufficientes juxta S. Alphonsum, adsint, scilicet, traditio mutua cum obligatione et possibilitate reddendi debitum, et vinculum indissolubile.

Si in tali muliere existat sensatio sexualis, pro ea remedium concupiscentiae habetur, sin minus, dantur saltem duo fines essentiales intrinseci matrimonii. Re quidem vera mulier carens ovariis et eodem tempore expers sensationis sexualis differt a muliere quae parturit sed sine sensatione sexuali est; nihilominus semper manent illi ovariis carenti duo fines essentiales intrinseci matrimonii.

Vetula communiter nequit parire et expers sensationis sexualis est, sed licite matrimonium contrahit. Conditio vetulae est eadem ac conditio mulieris ovariis orbatae. Lehmkuhl (Amer. Ecc. Rev. vol. 28, n. 3), in hanc assertionem urget excisionem ovariorum esse quid "positive actum contra primarium matrimonii finem," quum senectus conditio naturalis sit Chirurgus honestus aut inhonestus nunquam removet ovaria ut conceptio evitetur; operatio enim nimis periculosa est. Removentur ovaria primario ad morbum gravem medendum, et sterilitas consequens intenditur tantum per accidens. Remotio ovariorum igitur est quid per accidens actum contra "primarium finem matrimonii," (et iste non est finis essentialis intrinsecus) seu generationem prolis; et nihil refert etiamsi positive actum esset si non sit in fraudem legis. In casu mulieris habentis ovaria rudimentaria et nullam sensationem sexualem (casus enim quandoque contingit) quid sit "positive actum contra primarium finem matrimonii"? Estne una lex pro ista a natura castrata et alia pro muliere a chirurgo castrata et tertia pro vetula senectute castrata?

Inter ovaria et testiculos analogia est, etiamsi obstet D. Bossu, medicus Gallicus, a Professore Hild (Amer. Ecc. Rev. vol. 28, n. 1) et Eschbach citatus. Demptis ovariis sensatio sexualis destruitur haud secus ac quum testiculi removeantur, sed analogia incompleta est et claudit. Eunuchus insuper incapax est communiter intromissionis, et semper inseminationis. Moralistae vulgo docent inseminationem essentialem esse copulae carnali. Utcumque de inseminationis necessitate veritas sit (de qua infra) eunuchus nequit satisfacere primo matrimonii fini essentiali intrinseco, mulier expers ovariis satisdare potest.

Carentia aut occlusio tuborum Fallopianorum sterilitatem insanabilem efficit, sed usque adhuc nemo tenet illam carentiam vel occlusionem esse impotentiam. Conditio quoad potentiam generandi eadem est ac in carentia ovariorum, sensationem autem sexualem proprie carentia tuborum non efficit. Idem omnino dicendum est de carentia uteri. Si vagina remaneat capax talis mulier copulae carnalis est.

Mulier igitur impotens est sub lege ecclesiastica tantum in quinque casibus:

(1), ubi atresia vaginae aut adhaesio labiorum insanabiles sunt: haec atresia et adhaesio raro inveniuntur;

(2), in casu rarissimo in quo vagina in rectum aperit insanabiliter;

(3), ubi arctatio vaginae immedicabilis sit: haec rara est;

(4), quando adsunt tumores maligni aut incrementa morbida quae nequeant removeri;

(5), in casibus defeminationis aut Urningismi (de quibus infra). Sadismus et Masochismus et aliae perversitates sexuales ita infrequenter observantur, in gradu saltem in quo impotentiam creant, ut negligi possint.

II. Impotentia Maris. Tractus genitalis viri, a testiculis ad meatum urinarium seu orificium penis, ad centimetra 81 (uncias fere 32) extenditur. Ex testiculis chorda spermatica per inguen infra cutem transit, murum abdominalem penetrat per annulum inguinalem, et sub vesica urinaria urethram juxta cervicem vesicae intrat. Semen non ex uno fonte provenit. Secretio ista fluida est et cinerea, partim ex testiculis qui in scroto sunt oriens et partim ex vesiculis seminalibus, prostate, glandulis Cowperi, et folliculis cryptisque urethrae, quae omnia extra scrotum sunt. Elementum essentiale in semine cellulae sunt quae spermatozoa vocantur, et haec ex testiculis proveniunt. Locomotionis potentiam habent spermatozoa et in foecundatione ovum penetrant. Secretio glandularum quae spermatozoa fert alkalina est et removet aciditatem urinae; acidus enim spermatozoa destruit.

Erectio penis praecedit ejectionem seminis, et centra nervosa erectionis et ejectionis in chorda spinale apud lumbos sunt. Erectio effectus est dilatationis arteriarum penis et occlusionis venarum quae congestionem sanguinis efficiunt; postea musculi perineales aliique musculi erectionem perficiunt.

Impotentia maris in tria genera dividi potest; videlicet: Impotentia Psychica, Impotentia Atonica, Impotentia Organica:

I. Impotentia Psychica.

Species:

(1), Impotentia Psychica absoluta aut relativa;

(2), Sadismus;

(3), Masochismus;

(4), Fetichismus;

(5), Eviratio;

(6), Urningismus;

(7), Sodomia cum horrore;

(8), Gynandria;

(9), Metamorphosis Sexualis Paranoica;

(10), Anaesthesia Sexualis;

II. Impotentia Atonica.

Species:

(1), Impotentia Paralytica;

(2), Impotentia e venenis;

(3), Impotentia ex irritatione.

III. Impotentia Organica.

Species:

(1), absentia penis;

(2), penis multiplex;

(3), hypertrophia aut magnitudo penis vel praeputii;

(4), penis rudimentarius;

(5), adhaesio penis scroto, ingueni vel abdomini;

(6), hypospadias et epispadias;

(7), distortiones penis ex podagra, lue, rheumatismo, gonorrhoea;

(8), aneurysma corporum cavernosorum et varix venae dorsalis penis;

(9), frenum nimis curtum;

(10), anchylosis articulamenti coxendicis et abdomen permagnum;

(11), tumores et incrementa morbida circa genitalia, sicut herniae, hydrocele, haematocele, elephantiasis, lipoma, carcinoma, sarcoma, cystoma, enchondroma et fibroma;

(12), phthisis testiculorum et varicocele;

(13), anorchismus et castratio;

(14), prostratitis chronica;

(15), senectus;

(16), aspermia.

I. Impotentia Psychica.

1. Impotentia psychica ea est quae ex coercitatione inhibente cerebri in centrum genito-spinale exercitata devenit. "Impotentia ex maleficio" veterum moralistarum est. Timor, luctus, gaudium magnum, et aversio hanc impotentiam psychicam efficiunt. Quandoque nuper maritus propter excitationem passionis ejectionem praecocem vel erectionem debilem vel nullam habet. In his casibus impotentia temporanea est si medicus peritus prudensque sit. Haec impotentia relativa esse potest.

2. Sadismus. Haec paraesthesia sexualis et aliae perversitates ex excessu venereo deveniunt et impotentiam gignunt. Sadismus nomen habet a quodem libidinoso Gallico marchione de Sade, saeculo duodevicesimo vigente. Datur libido magna erga alium sexum sed cum crudelitate in objectum uriginis, quae crudelitas usque ad homicidium cum mutilatione frequenter extendit, vel saltem adesse debet humiliatio personae amatae.

Sadistae erant Nero et Tiberius; et exemplum infame erat Giles de Laval, qui A.D. 1440, supplicium capitis affectus est, post trucidationem Sadisticam fere 200 liberorum inter octo annos. Occisiones apud White Chapel Londini probabiliter Sadisticae fuerunt.

Sadista impotens est exceptis casibus in quibus crudelis vel saltem contumeliosus simul esse potest. Haec crudelitas gradus habet a Sadismo symbolico, in quo crudelitas simulatur, vel mere dramatica est, per contumeliam veram usque ad homicidium et anthropophagiam. Formae etiam bestiales et sodomiticae inveniuntur. Sadista nullo modo tanquam semper insanus considerari debet, species autem suae pessimae paranoicae sunt, et degeneratio neurotica frequens est in familiis Sadistarum.

Sadismus maris frequens est, sed Kraft-Ebing (Psycopathia Sexualis, Philadelphiae, 1893) tantum duos casus inter mulieres invenit. Fictiones antiquae de Lamiis et Marmolycibus ex actibus mulierum Sadisticarum ortae sunt.

Necrophilia, seu libido erga cadavera cum propensione ad mutilationem, species est Sadismi quae vulgo paranoica est.

3. Masochismus. Haec degeneratio nomen habet a fabularum scriptore Sacher-Masoch. Conditio est Sadismo contraria. Masochista uriginem habet magnam uti Sadista, et nulla datur potentia sexualis sine crudelitate vel humiliatione; crudelitas autem vel humiliatio in Masochistam ipsum vertenda est. Hic homicidium non intrat. Masochismus, natura sua passiva, vitium feminarum esse debet, sed solummodo casus unus in muliere

inventus est a Kraft-Ebing. Valde communis est inter mares.

Masochismus Larvatus est species hujus degenerationis in qua sordes physicae sordibus adduntur moralibus.

4. Masochismus Symbolicus seu Fetichismus. Fetichista potens est tantum praesente parte vestium, e. g., calccus mulieris, vel aliud objectum seu "Fetich," quodcumque sit. Aliquando mera imaginatio sufficit. Tonsores furtivi capillorum puellarum quandoque Fetichistae sunt, et in capillis longis virorum attractio sexualis quandoque est mulieribus. Virtus musicorum saepe in capillis Samsoniis est plus quam arte. Westermarck (History of Human Marriage. Neo-Eboraci. 1891) describit seditionem gravem mulierum in Madagascar quum Rex Radàma capillos longos militum tonderi jusserit. Relatio etiam est odores inter et passionem sexualem. Si nervi olfactorii catelli scindantur catellus nunquam canem femineam recognoscit. Meretrices gaudent odoramentis pungentibus, e.g., moscho.

5. Eviratio. Haec degeneratio gradus est Sodomiae. In gradu primo Sodomiae impotentia non necesse adest, in gradu autem secundo semper adest. In hoc secundo gradu homo Sodomiticus meretrix masculina evadit cum maribus, atque transformatio profunda et stabilis animi supervenit, ita ut mas se feminam esse in actu sexuali sentit. Hic est effeminatio usque ad statum criminalem pregrediens. Gradus initialis hujus status est amicitia inordinata inter duos pueros aut duas puellas.

In casibus firmatis evirationis, Sodomista agit tanquam pellex masculina aliis Sodomistis, aut in vestibus femineis ut uxor se gerit. Coloniae sunt Sodomistarum hujus generis in fere omne urbe magna; se invicem cognoscunt, societates, choreas publicas, et dialectum completam habent, praesertim Berolini, Lutetiae-Parisiorum, Neapolis et Washingtonii. Impotentes sunt ad copulam carnalem naturalem.

Sodomia cum effeminatione seu viraginitate frequens est inter mulieres. Hie degenerata marem mente evenit. Impotens vix dici potest, nisi propter aversionem sexualem a maribus.

6. Urningismus. Urning est vox Germanica ab Ulrichs inventa. Urningismus idem est in re ac Sodomia primi et secundi gradus, sed additur mollitia scriptionis poesis, ambulationum imminente luna, et aliorum hujusmodi.

7. Sodomia cum horrore est similis evirationi et defeminationi, sed vice frigiditatis adest horror alterius sexus.

8. Gynandria et Androgynia. In istis degenerationibus transformatio ita profunda est ut Sodomista masculinus circa pectus et in modo se gerendi feminae similis sit physice, et virago virum evadat aspectu. Exempla sunt bote et mujerado inter Sioux et Pueblos Indos Americanos.

9. Metamorphosis Sexualis Paranoica species insaniae est in qua patiens imaginat sexum suum mutatum esse. Insanabilis est.

Degenerationes istae fere omnes cum masturbatione incipiunt.

Medicatio moralis esse debet, sed Kraft-Ebing et von Schrenk-Notzing sanationem obtinuerunt ope hypnosis.

10. Anaesthesia Sexualis status est in quo vir aut mulier omnino caret sensatione sexuali. Illi secus habent corpora normalia. Conditio valde infrequens est. Kraft-Ebing enumerat decem casus congenitos in maribus, et duos in feminis. Anaesthesia sexualis acquisita etiam invenitur.

II. Impotentia Atonica.

1. Impotentia Paralytica. Centra nervosa sexualia in chorda spinale apud lumbos atonica esse possunt propter morbum generalem, aut venena, aut paralysem. Impotentia atonica frequens est in anaemia, diabete mellito, uraemia, cholaemia, lepra, rheumatismo, ataxia lumbali, lue chordae spinalis, myelitide, parese, et haemorrhagia cerebrali.

Aliqui tumores cerebrales impotentiam gignunt, paralysis diphtheritica causa est impotentiae temporaneae, et conditio invenitur in cachexia alicujus morbi tabescentis. Phthisicus autem saepe potens est usque ad finem vitae. Utrum impotentia perpetua sit necne ex natura morbi dependet.

2. Impotentia ex venenis. Potentia sexualis minuitur vel destruitur abusu venenorum vel absorbendo eadem, uti opium, morphina, chloral, potassii bromidum et iodidum, cannabis Indica, carbonei sulphidum, arsenium, antimonium, plumbum et iodum. Fabri, ut pictores (house-painters) et typographi, qui plumbo utuntur hoc modo patiuntur. Alcohol frequenter origo est impotentiae, et quandoque tabacum eumdem affectum habet sine alio indicio physico. Impotentia pervicax esse potest ex utraque causa.

3. Impotentia ex irritatione. Irritatio chronica urethrae prostaticae e libidine, gonorrhoea, masturbatione, urina acida aut neurosibus genito-urinariis impotentiam inducit. Neuroses centrorum sexualium sensibiles aut motoriae esse possunt, et neuroses motoriae quandoque impotentiam creant. Prognosis hujusmodi impotentiae fausta est nisi adsint spermatorrhoea et genitalium atrophia. Prostatorrhoea quoque impotentiae causa esse potest, et hic prognosis melior quam in spermatorrhoea est. Athletae, ut pugil, cursor, et alii ejusdem classis temporaliter impotentes esse possunt

III. Impotentia Organica. Mas impotens esse potest propter malformationes congenitas aut acquisitas, morbos et defectus genitalium.

1. Absentia congenita aut postnatalis penis impotentiam insanabilem creat.

2. Penis rudimentarius causa impotentiae est, sed casus amplificationis post matrimonium habentur.

3. Penis multiplex rarissime impotentiam creat.

4. Hypertrophia penis aut praeputii, et magnitudo relativa penis rarissime efficiunt hominem impotentem.

5. Adhaesio penis scroto, ingueni vel abdomini reddit hominem impotentem; sanabilis autem est conditio.

6. Impotentiam quandoque creat hypospadias, seu absentia partis inferioris urethrae, et epispadias, seu absentia partis dorsalis urethrae; sed conditiones a chirurgo sanari plerumque possunt, quandoque autem nequeunt.

7. Distortiones penis e podagra, lue, rheuraatismo, et gonorrhoea impotentem faciunt virum quandoque immedicabiliter.

8. Aneurysma corporum cavernosorum et varix venae dorsalis penis impotentiae a chirurgo sanabilis causae sunt.

9. Frenum glandulae penis curtum nimis incurvat penem, et ita vir impotens est. Conditio facile a chirurgo removeri potest.

10. Rarissime anchylosis articulamenti coxendicis et venter pinguedineus impotentem virum reddit. Anchylosis insanabilis est.

11. Tumores et incrementa morbida circa genitalia, ut herniae, hydrocele, haematocele, elephantiasis, lipoma, carcinoma, sarcoma, cystoma, enchondroma, et fibroma causae sunt impotentiae. Herniae, lipoma, hydrocele, haematocele, et quandoque elephantiasis scroti sanari possunt; tumores benigni et maligni aliquando removentur sed vulgo amputatione penis aut testiculorum.

12. Testiculi marcescunt in varicocele et impotentia sequitur. Varicocele amoveri potest, et si mature operetur chirurgus impotentia evitatur. Lues testiculorum destrui potest substantiam testiculorum nisi morbus mature sanetur, et impotentia potest esse absoluta. Tuberculosis testiculorum destruit organum.

13. Anorchismus seu absentia testiculorum bilateralis et congenita, idem efficit ac castratio. Conditio rara est Cryptorchismus, seu retentio testiculorum in abdomine, plerumque sterilitatem gignit non necessario impotentiam.

Castratio completa, morbo vel secus, impotentiae causa est, sed non necessario statim post castrationem. Gross (A Practical Treatise on Impotence, Philadelphiae. 1890) citat quattuor casus in quibus erectio permanebat, in uno homine usque ad decennium. Krügelstein (Henke's Zeitschrift. 1842. vol. I, p. 348) dicit virum quemdam post amissionem testiculorum uxorem foecundavit. Si casus revera contigerit, spermatazoa in vesciculis seminalibus permanserant.

14. Prostatitis chronica causa est impotentiae et plerumque insanabilis est.

15. Nulla regula firma dari potest de impotentia physiologica senectutis in maribus. Viri sexto et octogesimo anno non solum potentes inventi sunt sed etiam liberos generaverunt. Potentia generandi in maribus vulgo circa annum sexagesimum-secundum cessat, exceptiones autem multae inveniuntur. Potentia coeundi permanere potest longe post annum sexagesimum-secundum. Ecclesia igitur senes cujuscumque aetatis ad matrimonium admittit; et confessarius nihil rogat de potentia nisi ab ipso

sene interrogatur. Si confessarius sciat senem revera impotentem esse, non permittere debet matrimonium ejus.

16. Opinio videtur esse moralistarum ad habendam copulam carnalem necessariam esse non solum erectionem et intromissionem sed ettam inseminationem plus minusve perfectam.

Laymann (De Imped. Matr., cap. 11) ait: "Impotentia perpetua ad copulam perfectam dirimit matrimonium subsequens." Et addit: "Dixi perfectam, id est, quae fit cum effusione veri seminis in vas muliebre." Mastrius (loco jam citato) tenet inseminationem esse necessariam.

Lehmkuhl (Theol. Mor.) in definitione impotentiae absolutae dicit talem esse impotentiam quae "aliquem ad quamlibet copulam consummatam inhabilem reddit." Hic utitur vocibtis, copula consummata, in qua Ballerini requirit inseminationem quasi perfectam (op, cit., vol. 6, p. 178), et Sanchez (De Matr., lib. 2, disp. 21, n. 5) inseminationem imperfectam. Petrus de Ledesma quoque (apud Eschbach. Disputationes Physico-Theologiae. Disp. 200) de senibus loquens ait: "Si enim senes ita senio confecti et exhausti, quod nullo modo semine valeant, quamvis possint erigere membrum et penetrare vas, non possunt contrahere, et si contrahunt, matrimonium est invalidum."

Qui steriles sunt ita sunt ex tribus causis: (1) Aspermia, seu absentia absoluta seminis; (2) Azoospermia, seu absentia spermatozoon in semine; (3) Oligospermia, seu carentia alicujus partis seminis.

Aspermia vulgo efficitur occlusione urethrae vel obliteratione partis ejusdem. Defectus isti congeniti esse possunt vel ex morbo aut vulnere. Tumores tractus generativi claudere possunt aditum inter testiculos meatumque urinarium et ita aspermia efficitur. Spadones, etiamsi raro potentiam habeant intromissionis, aspermia afflictantur quoad partem seminis ex testiculis provenientem (in qua spermatozoa sunt), sed humor ex parte anteriori tractus generativi adesse aliquo modo potest.

Azoospermia fere eodem modo oritur, obstructio autem vel destructio prope testiculos est, et ita secretio aliarum partium tractus generativi exire potest, sed sine spermatozois.

In Oligospermia spermatozoa adesse possunt, sed quia aliud elementum seminis deest spermatozoa inertia sunt.

Casus inveniuntur in quibus propter malformationem semen perfectum in vesicam urinariam ejactatur.

Omnes istae conditiones insanabiles esse possunt, raro sanari possunt.

Estne vir aspermatosus, seu carens semine, impotens?

(1). Affirmant multi moralistae qui inseminationem requirant talem impotentem.

(2). Ejectio seminis confectio est actus sexualis viro, et alia in actu ejectioni mere conducunt.

In Azoospermia copula carnalis qua copula eadem est ac in coitu normali: microscopium solum aut sterilitas absentiam spermatozoon detegunt. In Oligospermia coitus quoque normalis esse paret.

Opinor virum aspermatosum impotentem sub lege esse quia nequit copulam sexualem perficere; e contra, virum oligo-spermatosum aut azoospermatosum steriles tantum esse.

Impotentia igitur definiri potest, in viro: Impotens est quum (1) vel absolute et perpetua, vel relative et perpetua, incapax sit intromissionis et quasi inseminationis; aut (2) quum spado sit, et ita sub decreto Sixti V veniat.

Impotentia in muliere definiri potest: quum nullam vaginam vel vaginam perpetuo impenetrabilem habeat; vel cum pathologice recuset marem.

Impotentia coeundi potest esse (1) aut antecedens aut consequens; prout matrimonii contractum anteit aut illi supervenit; (2) perpetua seu insanabilis, aut temporanea; (3) absoluta aut relativa, in quantum "aliquem ad quemlibet copulam consummatam inhabilem reddit, aut tantummodo usum matrimonii inter duas certas personas impossibilem facit" (Lehmkuhl).

Ut impotentia tanquam impedimentum matrimonii dirimens habeatur, debet esse: (1) impotentia coeundi; (2) insanabilis; (3) antecedens; (4) aut absoluta aut relativa.

AUGUSTINUS ÓMALLEY.

APPENDIX BLOODY SWEAT

The bloody sweat of our Lord mentioned in Saint Luke's Gospel (xxii., 44), has given rise to not a little discussion. The Greek text is:

The Vulgate has this text thus: "Et factus est sudor ejus sicut guttae sanguinis decurrentis in terram." The Douay version is: "And his sweat became as drops of blood trickling down upon the ground." The King James translation has it: "And his sweat was as it were great drops of blood falling down to the ground." The Greek text and the Vulgate and Douay versions are the same, but in the King James translation the words, "as it were great" differ somewhat from the statement in Greek.

The belief in the Catholic Church is that our Lord literally sweat blood through His unbroken skin, and this sweat is commonly deemed miraculous. Those that deny the sweat was really blood have no ground whatever for their assertion, because apart from all miracle bloody sweat can be a purely natural occurrence.

Dr. J. H. Pooley, in The Popular Science Monthly (vol. 26, p. 357), has an article on this subject in which he reported 47 cases of bloody sweat through unbroken skin. He, however, is of the opinion that our Lord's sweat had no real blood in it. Whatever his reason may be for this assertion he carefully conceals it.

Hemorrhage through the unbroken skin is a rare occurrence; but, as has been said, Dr. Pooley found 47 cases reported, and there are probably many others. The discharge may be pure blood which coagulates in crusts, or it may be blood mixed with sweat; it may be present over the whole surface of the body or only in those parts where the skin is thin and delicate. Commonly, bloody sweat is an oozing, but Hebra, is his Diseases of the Skin, tells of a young man that he himself observed, from whose legs and hand blood ran, sometimes in minute jets one-twelfth of an inch in height. The skin was sound, and the bloody sweat was not caused by any emotion.

The flow may be intermittent, appearing at intervals from a few hours to months. Sometimes the discharge is connected with skin diseases, but often the skin is unaffected. Examples have been found at every age and in both sexes, but this sweat is commoner in women. Du Gard reports an instance in a child only three months old, and Spolinus tells of such a sweat in a child twelve years of age.

Bloody sweat may occur in malaria; it may be connected with neurasthenic conditions, and it has been caused frequently by overwhelming emotion, as terror and anguish.

De Thou tells of a French officer who was in command at Monte Maro in Piedmont in 1552, who sweated blood after he had been threatened with an ignominious execution if he did not surrender the town. The same writer mentions a young Florentine, put to death in Rome during the pontificate of Sixtus V, who sweated blood before execution.

The Society of Arts at Haarlem reported the case of a Danish sailor who sweated blood through terror in a storm. This man was observed carefully by a physician on the ship. The physician at first thought the man had been wounded by a fall, but after wiping away the blood he discovered that the oozing came through uninjured skin. When the storm had ceased the sailor at once regained a healthy condition.

In the French Transactions médicales, for November, 1830, is narrated the case of a young woman who had turned from Protestantism to Catholicism, and after this conversion she grew hysterical because of persecution by her family. During the hysterical attacks she sweat blood from the surface of her cheeks and belly.

Before the Christian era bloody sweat was observed by Aristotle, Galen, Diodorus Siculus, and Lucan also mention such occurrences.

The stigmata of some saints are authenticated cases of bleeding through the sound skin of the hands, feet, and side during extraordinary sympathy with our Lord in His Passion, and deep mental concentration upon that Passion,—the stigmata of Saint Francis of Assisi, for example. Such bleeding is regarded in the Church as miraculous. Apart from any question of faith, there is no reason why they may not be miraculous, especially if the supernatural quality is supported by other facts; but, again, such stigmata can be natural. To prove, in general, that stigmata are miraculous requires commonly heroic sanctity as a background, and even then in all cases the proof is not necessarily absolute.

Focachon, a chemist at Charmes, applied postage stamps to the left shoulder of a hypnotised subject, and kept them in place with ordinary sticking plaster and a bandage. He suggested to the patient that he had applied a blister. The subject was watched, and after twenty-four hours the bandage, which had been untouched, was removed. The skin under the postage stamps was thickened, necrotic, of a yellowish-white colour, puffy with the serum of the blood and leucocytes, and surrounded by an intensely red zone of inflammation. Several physicians, including Beaunis, confirmed this observation; and Beaunis made photographs of the blister, which he showed to the Society of Physiological Psychology, June 29, 1885. (Animal Magnetism. Binet and Féré. New York, 1889.)

In Ricard's Journal de magnétisme animal, 2d year, 1840, pp. 18, 151, is a similar case. Prejalmini, in November, 1840, raised a blister on the healthy skin of a somnambulist by a piece of ordinary writing paper on which he had written a prescription for a blister.

At the meeting of the Société de biologie, on July 11, 1885, Bourru and Burot, professors of the Rochefort school, published records of epistaxis and of bloody sweat, produced by suggestion on a male hysteric. On one occasion, after the patient had been hypnotised, his name was traced with the end of a blunt probe on both the patient's forearms. There was, of course, no mark of any kind left on the arms. Then the patient was told: "This afternoon, at four o'clock, you will go to sleep, and blood will then issue from your arms on the lines which I have now traced." The man was paralytic and anaesthetic on the left side. He fell asleep at four o'clock, and while he was asleep the name appeared on the sound left arm, raised in a red wheal, and there were minute drops of complete blood (serum and corpuscles) in several places. There was no change on the paralysed right forearm. Later the patient himself commanded the arm to bleed and it did so. This second occurrence was observed by Mabille. (Binet and Féré. Op. cit., p. 199.) Charcot and his pupils at the Salpêtrière have often produced by suggestion alone the effects of burns upon the skin of hypnotised patients. The blisters in these cases did not appear at once but after some hours had elapsed. The blisters, of course, contained blood.

The weekly bleeding, through the unbroken skin, of the hands and feet of Louise Lateau is an example of stigmata in our own day, which may have been supernatural or natural. Physicians would call it natural, an effect of autohypnosis, but there is no reason why it may not have been just as miraculous as the stigmata of the saints. Professor Lefèvre of the University of Louvain, a physician, said her stigmata were miraculous. Theodore Schwann, the discoverer of the cell doctrine, deemed her condition natural.

In the Letters of the Rt. Rev. Casper Borgess, Bishop of Detroit, Michigan, is an account of a visit to Louise Lateau made in July, 1877. He says, "I first seated myself on the only chair in the room, which I had placed at the right side, near the head of the bed. Louise's two hands rested on several thicknesses of folded linen, spread over the bed-cover, and were covered with a folded linen cloth. This I removed. The hands were both heavily covered with blood; in some places it had congealed, and looked very dark; but in the centre, between the fore and little fingers, on the upper part of the hand, the blood was quite fresh and flowed freely. Not knowing at the time that the wiping of the hands causes her intense pain, I proceeded to wipe off the hands, for a more perfect inspection of the wound on each hand. The wound, or stigma, on the right hand seemed more than one inch in length, about half an inch at its greatest width, and was of oval shape. Turning the hand, I saw a wound of the same form in the palm of the hand, and opposite the wound on the back of the same. The blood seemed to rise in bubbles, forming in rapid succession, flowing in a spread stream down to the wrist. Examining the wound itself, I was well convinced that the skin of the hand was not broken nor in any way injured; and there was no sign of a wound made by any material instrument, sharp or dull. And, withal, the blood oozing out of the wound appeared a reality, and complete in form."

The bishop evidently uses the term "wound" in a figurative sense, because he draws attention to the fact that the skin was intact, continuous. She bled from the dorsal and palmar surfaces of the hands in areas shaped like the wounds represented by painters on the hands of our Lord. While the bishop was examining her hands Louise went into an ecstatic condition.

If the Church defines that a bloody sweat or the stigmata of a saint are supernatural, that definition, of course, ends the matter for Catholics as far as the particular case is concerned; but until such a decision has been made these conditions are all to be regarded as effects of natural causes working in a natural manner.

In many conditions where the nervous system can have influence a miracle is very difficult of proof from the context. There can, of course, be evident miracles in the cure of some nervous disorders, supposing the diagnosis to be certain. The sudden cure of advanced paresis would be as much a miracle as the sudden replacing of a lost femur. Commonly, however, in neuroses if there is an apparently miraculous healing or similar effect, the supernatural quality can not be established. Suppose Bernadette reported that she had seen the Blessed Virgin at Lourdes: the only safe thing to do in such a case is to deny the apparition until it has been proved. Suppose, secondly, that a patient who has been confined to bed for years by an hysterical paralysis, believed in the reality of the vision, had himself carried to Lourdes, and while at prayer there he suddenly stood up cured. That effect would prove neither the reality of the vision nor the supernatural quality of the cure; nor would it disprove either. We simply can not judge the case, because exactly the same effect has happened hundreds of times from purely natural impressions. If that same paralytic were lying in his bed at home and you set the house afire he would jump up and run.

If the patient, however, had been bedridden with a paralysis caused by certain degeneration of nervous tissue, and he were cured in the manner described, that effect would be supernatural, miraculous; always provided there is no error in the medical diagnosis.

There is a genuine diabetes and a pseudodiabetes. The latter condition may be diagnosed as true diabetes by a number of physicians, but it is only a symptom of hysteria. If the pseudodiabetes is suddenly cured, this cure may or may not be miraculous, but no one can say which is the truth; the probability is a hundred to one that the cure is altogether natural. There was a flourishing Christian Science congregation established in the west recently upon "miraculous" cure of a case of pseudodiabetes, which some ignorant physicians had called true diabetes, notwithstanding the fact that Christian Science does not believe in either diabetes or false diabetes.

We must not, then, call every strange event miraculous; nor, what is worse, are we rashly to make the supernatural a matter to be explained away loftily by the impudence of half science. A Belgian priest named Hahn wrote a monograph to the effect that the ecstatic conditions observed in the life of Saint Teresa were autohypnotic, and he succeeded in drawing upon himself the undivided attention of the Congregation of the Index and a serious disturbance of his peace of mind. He became a martyr to science. We all like to be "liberal," impartial; but from the religious Mugwump libera nos, Domine! Autohypnosis is always a mark of degeneracy in the natural order, and to call the ecstacy of a saint autohypnosis not only takes all worth from the manifestation, but the assertion is also untrue. There is a vast difference between the intense intellectual contemplation of a great saint in ecstacy, which leaves the person unconscious of the body and its surroundings, and the cataleptic trance of a neurotic patient who may mimic the saint.

Hypnotic or autohypnotic stigmata, and by stigmata here is meant bleeding from the hands, feet, and side, would be degeneracy of the mind and body in the natural order. Moreover, no clearly established cases are known, because conditions like those of Louise Lateau are by no means certainly physical from all points of view, as they would be if they occurred in an ordinary hysteric. In hypnosis or autohypnosis the subject's mind and body are degenerate; in sanctity, where at times may be displayed certain effects that resemble autohypnosis, there is always a sound mind. A saint may have an unsound, neurotic body, but a crazy "saint" or an hysterical "saint" is no better than any other lunatic or hysteric, and certainly anything but a saint. If a saint has stigmata, these external marks might come (1) miraculously, as a gratuitous sign of divine favour; (2) as an effect of natural, intense contemplation of the Passion of our Lord, producing these bleedings in a sound body; or (3) as an effect of a rational, intense contemplation of the same Passion, acting, more easily, on a neurotic body.

Scientific theorising on this matter is necessarily sterile, because such an investigation is only half material for science,—physical science. Science is not a bad thing in itself, especially when it minds its own business and keeps its place below stairs; but it never sympathises with sanctity, and there is no deep knowledge without sympathy. Fact-grinding made Darwin "nauseate Shakspere." Science can not see in the dark as genius and sanctity see, and if it does see in the dark it is no longer science but genius working on a scientific object. As Professor William James said: "Science taken in its essence should stand only for a method, and not for any special beliefs, yet, as habitually taken by its votaries, Science has come to be identified with a certain fixed general belief, the belief that the deeper order of Nature is mechanical exclusively, and that non-mechanical categories are irrational ways of conceiving and explaining even such a thing as human life." Science should recognise its own limitations and not meddle in attempted explanations of the inexplicable. Therefore, what of the stigmata of the saints from a scientific point of view? There is no scientific point of view.